W9-DFH-875

HIGH PERFORMANCE LIQUID CHROMATOGRAPHY IN ENZYMATIC ANALYSIS

HIGH PERFORMANCE LIQUID CHROMATOGRAPHY IN ENZYMATIC ANALYSIS

Applications to the Assay of Enzymatic Activity

Edward F. Rossomando
University of Connecticut Health Center

A Wiley-Interscience Publication
JOHN WILEY & SONS
New York / Chichester / Brisbane / Toronto / Singapore

Library of Congress Cataloging in Publication Data:

Rossomando, Edward F.

High performance liquid chromatography in enzymatic analysis.

"A Wiley-Interscience publication."

Bibliography: p.

Includes index.

1. Enzymes—Assaying. 2. High performance liquid chromatography. I. Title.

QP601.P783 1987 574.19′25 86–24690

ISBN 0–471–87959–2

Printed in the United States of America

10 9 8 7 6 5 4 3 2 1

To My Wife,
Nina,
And Our Children,
Natasha and Michelle,
With Affection

Preface

The importance of the introduction of high performance liquid chromatography (hplc) to studies in the life sciences is now widely recognized. Since its introduction, this method has been rapidly accepted by biochemists and more recently by biologists and clinicians. Such rapid acceptance should not be surprising, since advances in separation and analysis have usually been readily assimilated.

It is the ability of hplc to accomplish separations completely and rapidly that led to its original application to problems in the life sciences, particularly those related to purification. An analysis of the literature revealed that this technique was used primarily for the purification of small molecules, macromolecules such as peptides and proteins, and more recently antibodies. This application to purification has all but dominated the use of the method, and there has been a plethora of books, symposia, and conferences on the use of hplc for these purposes. However, it was only a matter of time before others began to look beyond and to explore the possibilities that result from the capacity to make separations quickly and efficiently.

What emerged from these early studies was the idea that hplc might be used as a method for the analysis of enzymatic activities rather than its traditional use as a tool for separation. This change in emphasis is particularly attractive to those who wish to make use of the activity of an enzyme as an indicator of cellular function, a determinant of a given stage of differentiation (or dedifferentiation), or even as a measure of gene function. In the past, because contaminating activities led to conflicting results, tedious purification of the enzyme was often necessary to clarify the results of ambiguous activity determinations brought about in part as a consequence of methods that measure only one component of the reaction mixture. Such ambiguous results will occur much less often with the hplc method, since its ability to separate quickly a group of related compounds allows for the assay of one activity in the presence of several others. Thus, the advantage of analyzing an activity after only a minimal amount of purification is inherent to the hplc technique.

This book describes the hplc method and explains and illustrates its use. Each chapter deals with a different aspect of the method, beginning with an overview and ending with a detailed summary. Throughout, an attempt has been made to focus on questions related to the assay of the activity of an enzyme rather than its purification. More detailed discussions on the theory of hplc and on its use for purification, particularly for the purification of proteins, will be found in the references at the end of each chapter.

No task of this magnitude can be completed without the guidance, inspiration, and help of many others. I wish to thank Jessica Hodge Jahngen, who introduced me to the possibilities and potential of hplc. She and E. G. Jahngen have provided much of the work in this volume from my laboratory. More recently, data and assistance have been provided by Jane Hadjimichael, whose persistence and insistence in the final stages helped it all come together. A special note of appreciation and thanks to Vickey Shockley for organizing and typing the manuscript together with Pamala Vachon, and to Sherry Perrie for the original artwork.

My appreciation is also extended to my colleagues Edward J. Kollar and James A. Yaeger for editing the early drafts of the manuscript. More recently, editing assistance has been provided by Cynthia Beeman, Mina Mina, and David Richards. Finally, I wish to thank Phyllis Brown for encouraging me to complete this task in the way I thought best and Stanley Kudzin for his interest in the subject and for providing the opportunity to write the book. Also, this work could not have been completed without the contributions of numerous investigators who consented to have their work cited here. My thanks to them as well.

EDWARD F. ROSSOMANDO

Farmington, Connecticut
January 1986

Contents

Chapter One

Application of HPLC to the Assay of Enzymatic Activities

Overview

In this chapter the anatomy of an enzyme assay will be described. The focus will be on the significance of separation and detection in the assay procedure. A classification of the methods used in the assay of enzymatic activities will be developed using the separation step as the criterion for the grouping. Having placed the high performance liquid chromatography (hplc) method within this classification, the question of when to use it will be examined, and some strategies developed for its use discussed. Those parts of the enzyme assay that will be affected by the selection of hplc as the method of choice will be analyzed.

1.1 INTRODUCTION

Increasingly, investigators in the life sciences have expressed interest in the application of hplc to the assay of enzymatic activities. Some of their reasons for considering hplc for this purpose are that it provides a method to enhance the separation of reaction components, allows extensive and complete analysis of the components in the reaction mixture during the reaction, can employ sensitive detectors, and can be used for purification.

A number of questions must first be addressed, however, concerning the biochemical reaction catalyzed by the enzyme, the assay conditions normally used for this enzyme, and the enzyme itself. This chapter has been designed to explore and answer these questions.

In Section 1.2 the anatomy of the enzymatic assay is presented, and from a dissection of its components it is possible to obtain an appreciation of how hplc can be used. In Section 1.3, a classification of enzyme methods is developed that allows the advantages and limitations of the hplc method to be presented fairly. In Section 1.4, criteria are developed for the selection of hplc as an assay system. Wherever possible, these points will be illustrated with examples taken from work carried out in the author's laboratory.

1.2 ANATOMY OF AN ENZYME ASSAY

The assay of an enzymatic activity is composed of several discrete steps or events (Fig. 1.1). The first is *preparation* of both the reaction mixture and the enzyme. The reaction mixture usually contains such components as the buffer used to establish the correct pH, the substrate, and any cofactors such as metals that may be required for catalysis. Preparation of the reaction mixtures involves mixing these ingredients in a reaction vessel such as a test tube or, for some assay methods, a cuvette. In some cases the reaction mixture is brought to the required temperature prior to initiation of the reaction. The enzyme must also be prepared. This, however, is a more complex topic and will be discussed in detail in a later chapter.

In most cases, the second step in the assay is *initiation* and *incubation*. A reaction can be initiated by the addition of the enzyme preparation to the substrate in the reaction mixture or vice versa. This step is considered the start of the reaction, and all subsequent time points are related to this time.

Many reactions require *termination*, which is the step that brings about the cessation of catalysis and thus stops the reaction. Termination may be achieved in several ways, all of which usually involve inactivation of the enzyme.

Termination is often followed by *separation* of the components in the reaction mixture. Most often separation involves isolating the substrate from the reaction product.

Detection, the fifth step, refers to that process by which the amount of

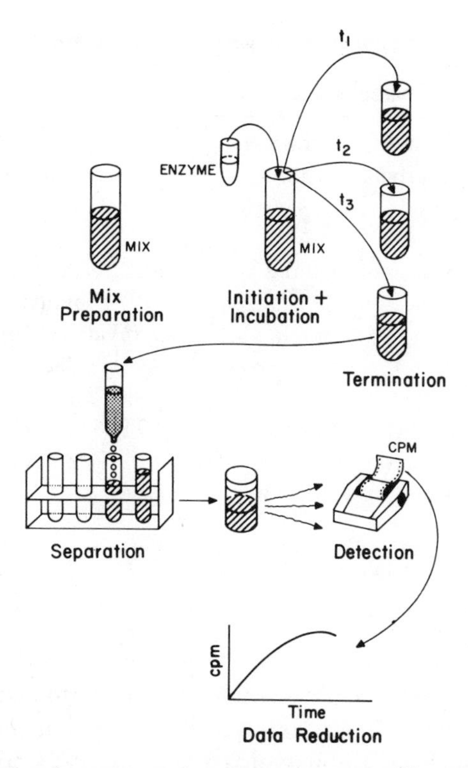

Figure 1.1 Schematic of a representative enzymatic assay to illustrate its several components. The reaction mixture is prepared (Mix Preparation) and the reaction can be started (Initiation) by the addition of the enzyme. During the reaction (Incubation), samples are removed at intervals labeled t_1, t_2, and t_3, and the reaction is stopped (Termination) by inactivating the enzyme. The incubation mixture is fractionated (in the illustration a traditional chromatographic column is being used), and the product is isolated from the substrate (Separation). In the assay illustrated, a radiochemical has been used as the substrate and therefore the amount of product that formed is determined by its collection, the addition of scintillation fluid, and the measurement of radioactivity by scintillation counting (Detection). The progress of the reaction is given by the amount of radioactive product recovered (Data Reduction).

product formed by the enzyme during a specific incubation interval is determined.

Finally, the last step in an assay involves *reduction* of the data. This step includes all procedures in which the data are analyzed and graphed to determine initial rates as well as kinetic constants.

Table 1.1 Classification System For Enzymatic Assay Methods

Assay Method	Characteristics	Example
Continuous	Separation of substrate(s) from product(s) not required	$4NP \rightarrow 4N + P_i$ colorless yellow
Coupled	Separation not required for detection	$PEP + ADP \rightarrow$ pyruvate + ATP pyruvate + NADH $\rightarrow$ lactate + NAD
Discontinuous	System for separation of substrate(s) from product(s) required for detection	$ATP + AA \xrightarrow{Enz} Enz-AA-AMP + PP_i$ $Enz-AA-AMP + tRNA \rightarrow tRNA-AA + AMP$

Not all steps are involved in all assay methods, and in some methods one or more of the steps might be complex. The introduction of hplc as an enzymatic assay method has improved the separation and detection steps primarily, although its use may also affect the preparation and termination steps.

1.3 CLASSIFICATION OF ENZYMATIC ASSAY METHODS

The methods in use for the assay of enzymatic activities may be divided into three groups. These will be referred to as (1) continuous, (2) coupled, and (3) discontinuous methods (see Table 1.1).

1.3.1 Continuous Methods

Continuous methods do not require a separation step prior to detection. For assays using this method, the substrate and product must differ in some property such that either one may be measured directly in the incubation solution. For example, the activity of an enzyme catalyzes the conversion of 4-nitro phenyl phosphate (4NP), a colorless compound, to

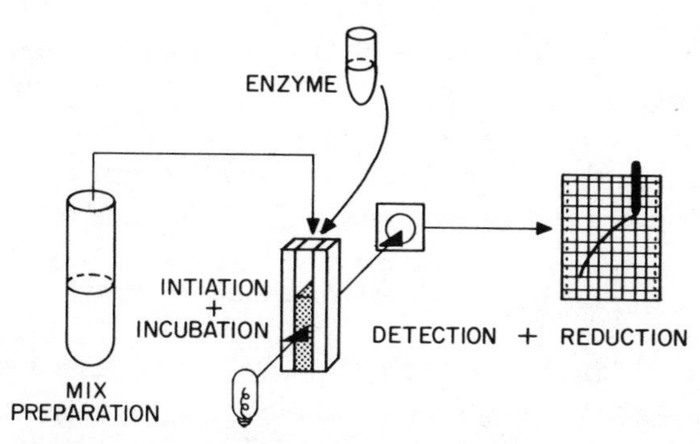

Figure 1.2 The assay of an enzymatic activity by the continuous assay method. In the illustration, the reaction mixture is transferred to a cuvette, which is shown in place in the light path of the spectrometer. The addition of the enzyme directly to the cuvette initiates the reaction. Product formation results in a change in absorbance, which is monitored continuously by the detector. This change signals a deflection on a recorder. Note that product formation requires neither termination of the reaction nor separation of the substrate from the product.

4-nitrophenol, which is yellow and has an absorption maximum at 510 nm. Since the substrate does not absorb in this region of the spectrum, the reaction can be carried out directly in a cuvette (Fig. 1.2), and the amount of product formed may be determined continuously by measuring the change in optical density with time at this wavelength.

1.3.2 Coupled Method

In the second category of assays, the coupled assay method, activity is measured indirectly. In this method two reactions are involved: The first, the reaction of interest, such as A→B and the second, the reaction that converts B to C and might be referred to an an *indicator reaction*, not only because it uses the product of the first reaction, that is, B, as a substrate, but also because the formation of C may be assayed by a continuous method, that is, without a separation step. In this way, the two reactions are coupled, the product of the first reaction, B, acting as the substrate for the second reaction.

For example, pyruvate kinase may be assayed by such a method. This enzyme catalyzes the reaction

$$\text{Phosphoenolypyruvate (PEP)} + \text{ADP} \rightarrow \text{pyruvate} + \text{ATP}$$

This, of course, is the reaction of interest that cannot be assayed directly by the continuous method. However, when a second enzyme, a dehydrogenase, such as lactate dehydrogenase, is added as the indicator together with pyruvate and NADH to the reaction mixture, a second reaction occurs and NAD forms in the cuvette as follows:

$$\text{Pyruvate} + \text{NADH} \rightarrow \text{lactate} + \text{NAD}$$

The formation of NAD may be followed in a continuous manner by the decrease in absorbance at 340 nm, and therefore the progress of the kinase reaction of interest may be followed through this coupling of the formation of pyruvate to the formation of NAD.

1.3.3 Discontinuous Method

In the discontinuous method, product must be separated from the substrate in order to measure activity. Assays characteristic of this group usually require two steps, since separation often does not include detection. Thus, first, the substrate and the product are separated, and usually the amount of product formed is measured. Assays that use radiochemical substrates are included in this group, since radiochemical detectors are unable to differentiate between the radiolabel of the substrate and that of the product. Examples of enzymes whose assay methods fall into this category are legion and are characterized by a separation step.

As an illustration, consider the assay to measure the activity of the tRNA synthetases. These enzymes catalyze the covalent attachment of an amino acid, usually radioactive (as indicated by the asterisk in the reaction), to tRNA as follows:

$$\text{ATP} + {}^*\text{AA} + \text{Enz} \longrightarrow \text{Enz-AMP-}{}^*\text{AA} + \text{PP}_i \quad (1)$$
$$\text{Enz-AMP-}{}^*\text{AA} + \text{RNA} \longrightarrow \text{RNA-}{}^*\text{AA} + \text{AMP} \quad (2)$$

The activity is usually followed by measuring the amount of RNA-*AA, the product of reaction (2) formed during the incubation. Since the radiochemical detector cannot differentiate the free radioactive amino acid used as the substrate from that bound covalently to the RNA the free and the bound amino acids must be separated prior to the detection or quantitation step.

This separation step requires first the addition of an acid such as trichloroacetic acid (TCA) to the sample, which also serves to terminate the

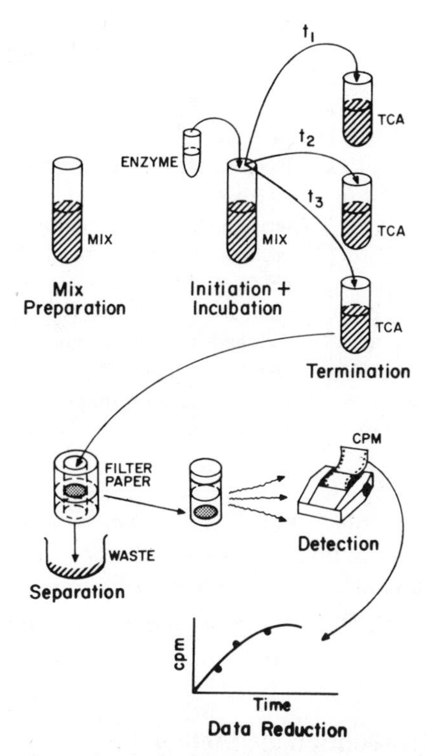

Figure 1.3 The assay of an enzymatic activity by the discontinuous assay method. In assays of this type, the reaction mixture is prepared, and usually the reaction is started by the addition of the enzyme. Samples are removed at intervals t_1, t_2, t_3, and the reaction is terminated by transferring the sample to a solution that inactivates the enzyme. In this illustration, a radioactive substrate is converted to an acid-precipitable radioactive product (precipitable in trichloroacetic acid, TCA) while the substrate remains soluble in the acid. Thus two components can be separated by filtration. The product is shown being collected on the filter while the unreacted substrate flows through into the filtrate. The amount of radioactivity trapped on the filter is determined by scintillation counting. When these data are graphed as a function of sample time, i.e., t_1, t_2, and t_3, they provide the kinetics of product formation.

enzymatic reaction. However, since TCA also precipitates the RNA and any radioactive amino acid covalently linked to it, the reaction product RNA-*AA will be precipitated as well. And since the precipitate can be separated from the soluble components by a sample filtration step, the separation of the bound from the free amino acid can be accomplished. As illustrated in Fig. 1.3, the reaction product, which is trapped on the filter as a precipitate, can be detected by transferring the filter to a scintil-

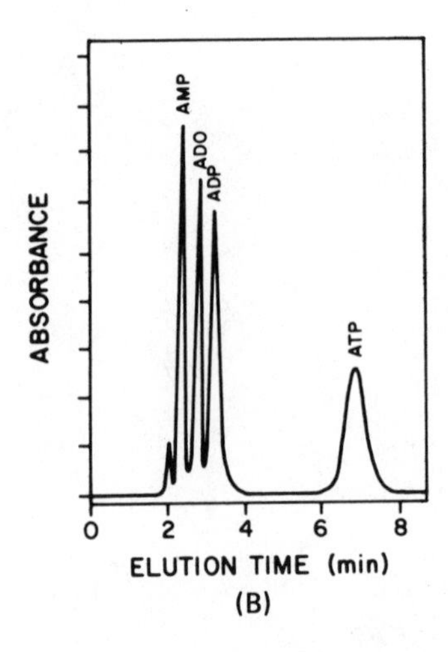

Figure 1.4 Separation of substrates and products of an adenosine kinase reaction on ion-pair reverse-phase hplc. The separation was carried out on a prepacked C-18 (μBondapak) column with a mobile phase of 65 m*M* potassium phosphate (pH 3.7) containing 1 m*M* tetrabutylammonium phosphate and 5% methanol. The column was eluted isocratically, and the detection was at 254 nm. The relative elution positions (elution times) of ATP, ADP, ADO, and AMP are shown.

lation counter for quantitation and, of course, measuring the amount of product formed. It should be noted that since assays of this design usually focus on one component at a time, no information is obtained about the amount of ATP, AMP, PP_i, or free amino acid during the course of the reaction.

1.3.4 HPLC as a Discontinuous Method

Within the framework of the scheme described above, the hplc method would be classified as discontinuous, since a separation step is part of the procedure. However, because termination can be accomplished by injecting the sample directly onto the column, the hplc detection is usually "on-line," that is, carried out continuously with separation. Thus, the separation and detection steps merge into a single operation, which for all practical considerations is a "continuous" method.

In addition it should be noted that unlike many other discontinuous assays that focus on only one of the components of the reaction, the hplc assay offers the potential to monitor several. For example, consider adenosine kinase, the enzyme that uses two substrates and forms two products according to the reaction Ado + ATP→AMP + ADP. Since hplc can readily separate all four compounds (see Fig. 1.4), and all four compounds can be detected at 254 nm, it is apparent that, with the hplc method, the level of each component can be monitored during the course of the reaction, providing a complete analysis of each "time point."

Having a complete analysis of the contents of the reaction vessel during the incubation can be helpful in another way; it provides information on what is not present as well, and since most other assay methods are designed to detect only one component, it is often difficult to account for an unexpected result occurring during a study. For example, consider the results obtained during the purification of the enzyme E-1, which catalyzes the conversion of substrate A to product B. Consider also that the method used to follow activity measures only the amount of B in the incubation mixture. As illustrated in panel I of Fig. 1.5, when the activity E-1 is assayed in the crude sample, the formation of substantial product (B) is observed (graph line 1).

Imagine now that the enzyme sample is purified further and the now purified enzyme is assayed for the same activity E-1 by the same method. However, in this case, following the addition of the substrate (A), the formation of product (B) during the course of the reaction is greatly reduced (graph line 2). While this result might be explained by a true loss of E-1 activity, it might also be a result of an increase in the activity of E-2, a second enzyme which catalyzes the degradation of B to C (Fig. 1.5, panel II). In the absence of data on the level of the substrate (A) during the course of these reactions, this possibility cannot be excluded.

To test this possibility will require an assay for E-2, which in turn requires a method for the measurement of the amount of C. Therefore, a reaction mixture optimized for E-2 will have to be prepared, and determinations of C formation in both the crude and purified samples will have to be carried out. If the data obtained in these determinations appeared as shown in panel II of Fig. 1.5, then these results would show that in the crude sample E-2 activity was indeed lower than that observed in the purified sample and would indicate that the purification of E-2 could account for the loss of activity of E-1 during the purification. This example clearly shows that an assay method that measures only the levels of a single compound such as the product can provide only very limited results.

In contrast, with a well-designed hplc assay for activity E-1, capable of separating A, B, and C, the levels of each may be obtained from an analysis of a single sample from the reaction mixture of both the crude and purified samples (panel III of Fig. 1.5). In fact, the data obtained provide information not only on what compounds are present but also on what are not. The availability of such negative information can provide the "data"

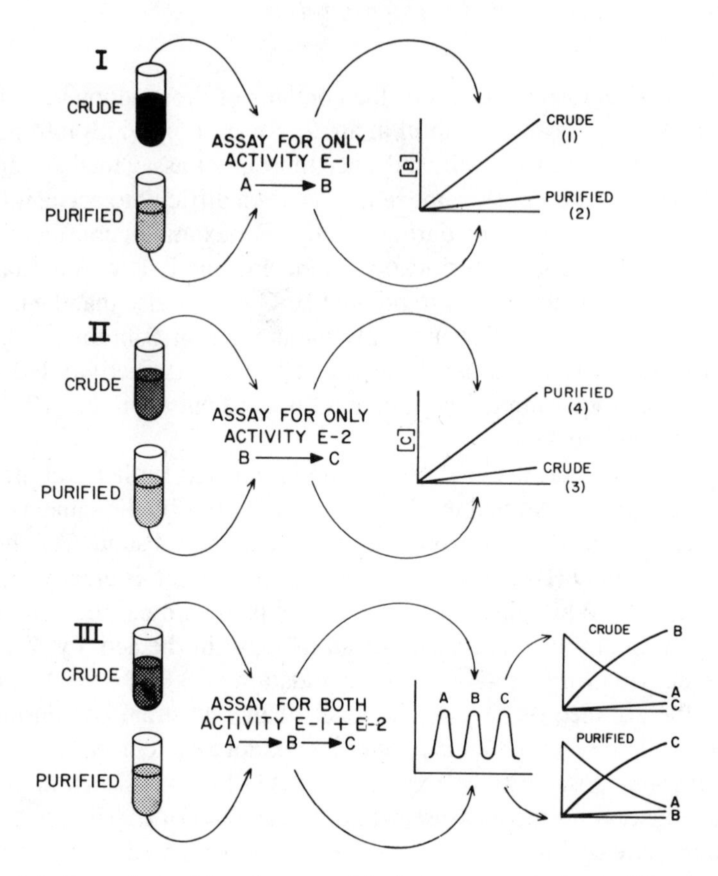

Figure 1.5 Comparison of the advantage of the hplc assay method to traditional methods of assay when following the activity of an enzyme during purification. The assay of the hypothetical enzyme, called E-1, which catalyzes the conversion of the substrate A to the product B, is depicted in panel I. In this traditional method, an assay would be developed to follow the production of the product B. Note that in the crude extract E-1 produced B at the rate shown by line (1). However, after purification, E-1 produced B at a much slower rate, as indicated by line (2). To understand the reduction in rate, a second hypothetical enzyme was proposed that would convert B, the product of E-1, to a new product C. However, another traditional assay was used, and only the formation of C was measured. As illustrated in panel II, E-2 activity was measured in both fractions, and while activity was found in both the crude and purified fractions, the rate of C formation in the purified fraction shown by line (4) was significantly greater than in the crude as shown by line (3). In panel III, the activity of the same two enzymes is shown being measured by the hplc method. In this case, the assay was developed to separate A, B, and C simultaneously, and therefore it was possible to measure the activity of both E-1 and E-2 simultaneously. In panel III, the results of assays carried out on both the crude and the purified preparations are shown. The difference in levels of A, B, and C during the course of the incubation with both the crude and purified preparations is illustrated.

Table 1.2 Questions to be Considered Prior to the Selection of HPLC for the Assay of an Enzymatic Activity

1. *Separation and Detection*
   Must product be separated from substrate for analysis?
   Are detectors available?

2. *The Reaction Mixture*
   Are there limits to the total volume of the incubation mixture?
   Are cofactors such as metals a problem?
   How will the reaction be terminated?

3. *The Enzyme*
   Is the enzyme pure, or will contaminating activities be present and affect product levels?

to exclude alternative explanations that were proposed to explain the unexpected results.

1.4 CRITERIA FOR THE SELECTION OF AN ASSAY METHOD

The hplc assay may not always be the procedure of choice, and several points should be considered before deciding. These points are summarized in Table 1.2.

1.4.1 Separation and Detection of Components

To utilize hplc for an enzymatic assay requires first a system for separating the components. This involves the selection of (1) a solid phase (the column packing), (2) a mobile phase (the column eluent), and (3) a method of elution from the solid phase by the mobile phase. Two procedures are generally used for elution: the isocratic and the gradient methods. In the isocratic method the mobile phase composition remains constant throughout elution, while in the gradient method the mobile phase varies in some parameter and in some fixed manner such as a linear increase in salt concentration during elution. In addition, the hplc method requires a monitor for the detection of the product. While a variety of detectors are available for monitoring various properties of molecules, a number of compounds cannot be detected on-line at present.

1.4.2 The Reaction Mixture

An important consideration with regard to the hplc assay method is whether the reaction conditions previously developed for the assay of this activity can be adapted for use with hplc assay. For example, is the reaction mixture of sufficient volume to permit the withdrawal of multiple samples? For those assays carried out in volumes of a few microliters, it is virtually impossible to withdraw samples with sufficient volume for analysis on the hplc system. Thus, unless dilutions can be made after sampling, the hplc system would have to be ruled out in such cases.

Other factors should be considered as well. These include whether or not there are components in the reaction mixture that might make using hplc difficult. Such components include metals, which often produce difficulties in the interpretation of chromatograms. While the problem of metals can be solved easily by the addition of chelators, for other problems the solution may be more complex. For example, in those cases where the reaction must be terminated prior to injection, the termination process itself often alters the incubation mixture. Termination by acids, such as trichloroacetic or perchloric acid, reduces the pH of the sample. Since differences between the pH of the sample and the mobile phase can produce discrepancies in chromatographic profiles, the reduction in pH brought about by the termination may cause problems in interpretation. Also, termination increases the possibility of producing a precipitate, which will have to be removed before the sample can be injected into the column, to prevent clogging of the system. While the removal of precipitates is not a difficult task, it does introduce an additional step into the assay procedure.

1.4.3 The Enzyme Sample

A final point to be considered in the use of hplc as an assay procedure is the enzyme itself. Will the activity be a pure enzyme; or will it be part of a rather crude cell-free extract, or will it be present in a fermentation broth? In the case of the latter two samples, the presence of contaminating activities must be considered. While as mentioned above, these activities, by affecting the recovery of the product or even by affecting substrate levels during the course of the reaction, could easily cause problems with other assay procedures they are not a problem for the hplc assay method. Thus, hplc should be considered first when assaying an activity in some crude extracts.

1.5 SUMMARY AND CONCLUSIONS

The assay of the activity of an enzyme can be subdivided into several steps. These include the formation of a reaction mixture, preparation of an enzyme sample, the combination of the two to initiate the reaction, incubation of the reaction, termination of catalysis, separation of components, their detection, and finally, reduction or processing of the data.

Not all assays require a separation step, and this fact may be used to develop a classification scheme for assay methods. Those assays which require no separation have been grouped under the heading "continuous assay method," while the "discontinuous method" describes those which do.

The necessity for use of a discontinuous assay method does not automatically mean that the hplc method is the procedure of choice. For hplc to be suitable, it must be possible to separate the components, and some method for detection and quantitation must be available. Next, neither the ingredients in the reaction mixture nor those used to terminate the reaction should produce problems for the separation and detection. Finally, the enzyme itself should be considered. Excess protein can contaminate columns, and extraneous enzymes can cause problems in quantitation.

GENERAL REFERENCES

Reviews of Liquid Chromatography

Ettre, L. S., The evolution of modern liquid chromatography, *LC Mag*. ***1:***108 (1983).

Freeman, D. H., Liquid chromatography in 1982, *Science* ***218:***235 (1982).

Classification of Enzyme Assay Methods

Dixon, M., Webb, E. C., Thorne, C. J. R., and Tipton, K. F., Enzyme techniques, *Enzymes*, 3rd Ed., Academic, New York, 1979, Ch. 2.

Chapter Two

Concepts and Principles of High Performance Liquid Chromatography

Overview

In this chapter some of the basic concepts and principles of liquid chromatography will be introduced. Some background on the development of high performance liquid chromatography (hplc) will be given, and the basic components will be described briefly.

The chromatogram will be introduced as the record of the separation. The information it does and does not contain will be discussed. Examples of different chromatographic profiles will be presented and interpreted.

A strategy for the selection of the stationary phase will be developed based on a discussion of the mechanism underlying the separation involved in gel filtration, reverse-phase, and ion-exchange chromatography.

The mobile phase will be considered, including its composition, preparation, and use.

The problem of column maintenance, particularly when the column is used for enzyme assays, will be discussed, and cleaning solutions recommended, and a method for monitoring column performances described.

2.1 INTRODUCTION

Chromatography, the separation of classes or groups of molecules, in principle requires two phases. In liquid chromatography one of the phases is liquid and the other a stationary phase, often bonded to a solid. In days gone by, a stationary phase widely used in biochemistry laboratories was potato starch, and graduate students often found themselves up to their elbows in white "gooey stuff" making large batches of starch. After mixing, the starch would be poured into rectangular forms where it hardened for later use as the solid phase for the chromatography.

The sample was applied to one end of the block (or column if the stationary phase was poured into a vertical cylinder) and eluted from the other end (the bottom of the column) by allowing a mobile phase or solvent to flow through the stationary phase. Since the molecules of the sample are carried along by the mobile phase, the time required for a group of molecules to emerge from the stationary phase, other things being equal, was a property of the packing material. Those that emerged in the shortest time were considered not to have been affected by or to have interacted with the stationary phase; whereas those that emerged later did interact. Emergent time—or, as it is more often called, elution time—could be affected by two parameters: the distance, that is, the path, that the molecules follow as they travel through the packing and the rate (velocity) at which they traverse the packing. These two parameters may be expressed in the relationship

$$\text{Emergent time} = \frac{\text{distance traveled}}{\text{rate of travel}}$$

2.2 THE INTRODUCTION OF HPLC

It has been known since the early days of liquid chromatography that the size of the particle used for the stationary phase affected the separation, or resolution, in a rather direct way: the smaller the particle, the better the separation.

However, with columns that used gravity to pull the mobile phase around the particles, a lower limit to particle size was reached, since the smaller the particles the tighter they packed, eventually cutting off the flow of solvent. Thus, a need developed for a pressurized solvent delivery system able to pump the mobile phase through the packing. Of course, as such pumps were developed, the old packings were found to collapse with the increased pressure and new packing materials were required that could withstand these pressures. Together, pumps and new packing materials provided resolution and separation not achieved previously with other methods. As a fringe benefit, the new technique has considerably shortened the time required to carry out separations. Separations previously taking hours or even days are now accomplished in minutes.

Another fringe benefit is the increased sensitivity provided by the detectors. Thus, it is no longer necessary to resort to the old ploy that I was shown rather jokingly by a fellow student during my graduate school days when I was examining the results of an experiment in which radiolabeled compounds had been used. I remarked to him on the incredibly low number of counts obtained during the experiments and expressed the wish for more counts. This prompted my fellow student to turn to the counter and change the counting-time dial from 1 min to 10 min. Needless to say, a recount of the same samples produced more counts. In hplc it is not necessary to resort to such tactics to obtain increased sensitivity. It is a result in part of the geometry and low volume of the flow cells used in the detectors.

2.3 BASIC COMPONENTS AND OPERATION

An hplc system, shown schematically in Fig. 2.1, consists of (1) a *solvent reservoir*, which contains the eluent or mobile phase; (2) a *pump*, often called a solvent delivery system; (3) an *injector* through which the sample is introduced into the system without a drop in pressure or change in flow rate; (4) the *analytical column*, which is usually stainless steel and contains the solid packing or stationary phase; and (5) a suitable *detector* to monitor the eluent.

Also shown in Fig. 2.1 are two other components: a precolumn and a guard column. The *precolumn*, located in the system between the solvent reservoir and the pump, acts to filter out any impurities in the mobile phase before they reach the pump heads and the analytical column. In the case of analytical columns made of silica, impurities in the solvents can result in leaching of the silica. If solvents are made on a daily basis and are filtered (see below), the precolumn may not be needed.

The *guard column*, by virtue of its location between the injector and the analytical column, functions to remove any insoluble material and

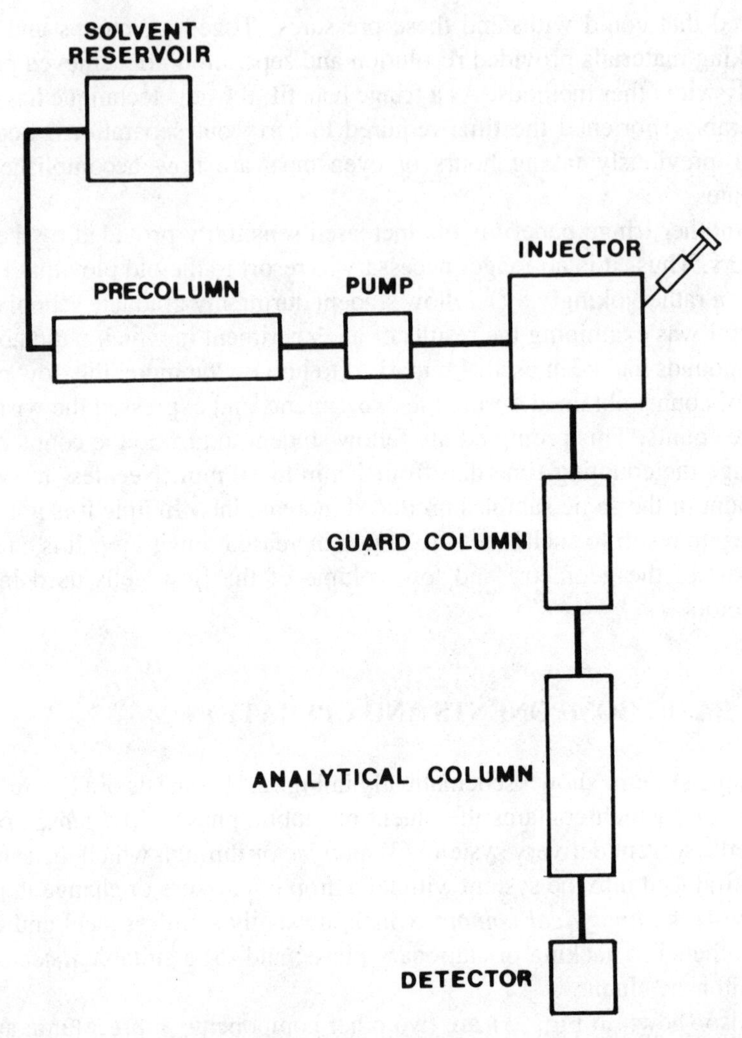

Figure 2.1 A representative diagram of the components of an hplc system. Solvent flow is from top to bottom. The size of each box in the diagram should not be taken as the actual size of each component.

other debris that might have been injected and would otherwise clog the analytical column. For example, when used with enzyme assays, the guard column will remove precipitated proteins or other insoluble material carried over from the incubation mixtures. Guard columns can themselves get clogged, and as a consequence, if the pump is to maintain a constant flow rate it must generate greater pressures to drive the solvent through the clogged filter. This increase in pressure, referred to as "back pressure," can be eliminated by cleaning and repacking the guard column.

To operate an hplc system, the sample is introduced through the injector into the system and is then pushed through the analytical column by the constant pumping of solvent (or mobile phase) from the reservoir through the system (Fig. 2.1). The mobile phase can be delivered in two ways: *isocratically*, that is, at constant composition, or in the form of a *gradient*, when the composition is varied. How to decide which form to use is described in Chapter Three. Injectors are also described in more detail in Chapter Three. Additional information about the operation of pumps can be obtained by consulting the references.

Following its emergence from the other end of the column, the eluent flows through the detector. Detectors operate on various principles. For example, some monitor the ultraviolet, visible, or fluorometric properties of molecules; others monitor radioactivity; and still others monitor differences in oxidation-reduction potential and refractive index. These detectors are listed in Table 2.1 together with some examples of the specific reactions with which they have been used.

2.4 COUPLING THE COMPONENTS: ON THE PERILS OF FERRULES

While one of the most confusing steps for the new user of hplc is deciding what equipment to purchase, an even more difficult and frustrating step occurs after the equipment has arrived and connections must be made between the solvent delivery system (the pump), the various columns, and the detector, for connections are not made with the more familiar easy-to-use flexible plastic, but with unfamiliar stainless steel tubing. Since stainless steel cannot be cut with scissors or easily coupled with plastic connectors, special tools must be used for cutting and connections must be

Table 2.1 Detectors and Their Applications

Detector	Reaction Analyzed
1. UV spectrometer	ATP → ADP + P_i IMP → Ino + P_i
2. Fluorometer	FoTP* → cFoMP cFoMP → FoMP
3. Radiochemical	Hyp + PRPP → IMP + PP_i AMP + ATP → 2ADP
4. Electrochemical	L-Dopa → dopamine L-5-Hydroxytryptophan → serotonin
5. Refractive index	Maltoheptose → oligosaccharides

*FoTP; formycin 5-triphosphate.

made with nuts and bolts and fittings called ferrules. Unfortunately, ferrules are not all the same, and once in place they are not easily removed.

A schematic of a ferrule is shown in Fig. 2.2. It is swaged, that is, attached, when the end of one piece of stainless steel tubing is coupled to another. The coupling itself involves a bolt (B) and a nut (N), which are assembled as follows: The bolt is placed on the tubing, followed by the ferrule, and the nut is threaded to the male bolt, trapping the ferrule between and thus swaging the ferrule to the tubing. The appearance of the ferrule on the tubing before compression is shown in the upper panel of Fig. 2.2, and its appearance within the fitting after compression in shown in the lower panel. It should be noted that to avoid damage to the ferrules, they should not be overtightened, and it is also important to keep in mind that the connecting tubing should be short and the connectors few to avoid excess mixing space.

2.5 THE CHROMATOGRAM

Information about the separation is displayed on a chromatogram, which is obtained by converting the detector output to an electrical signal and

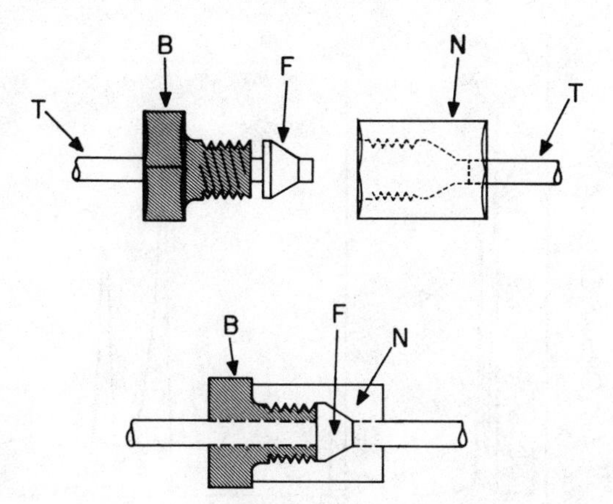

Figure 2.2 Cross-sectional diagram of the components used to couple two pieces of stainless steel tubing. Top: Units before swaging. T, tubing to be joined; B, "male" bolt; F, ferrule; N, "female" nut. Bottom: After swaging. The pressure of the bolt on the nut has forced the ferrule to seal the joint between the two ends of the tubing.

following this signal on a recorder as a function of the time after the loading of the sample. Figure 2.3 shows a representative hplc chromatogram of a sample containing two species of compounds denoted A and B. In this example, while both enter the column at the same time (with the injection of the sample), compound B traverses the column at a faster rate than compound A. As shown in Fig. 2.3, compound B will emerge and be

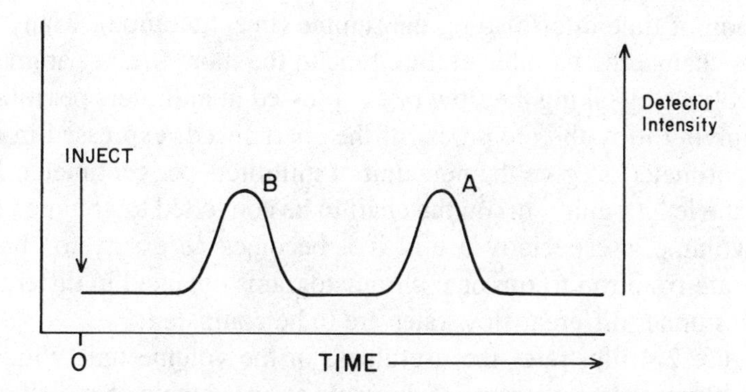

Figure 2.3 A representative hplc chromatogram. The separation of two compounds, A and B, is shown. The time of injection is taken as zero time, and the elution position is shown as a function of time after injection. The amount of each of the compounds A and B in the original sample is given by the peak height or area as represented on the tracing.

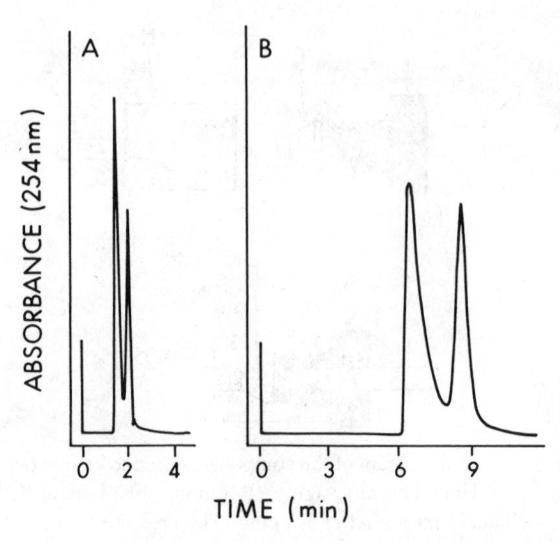

Figure 2.4 A comparison of the chromatography of the same two compounds carried out at flow rates of (A) 1 mL/min; (B) 2 mL/min.

detected first and A later. The time of injection of the sample, marked on the chromatogram by the arrow, is taken as zero time, and the time after injection is determined from the speed of the recorder. Of course, the rate of fluid flow is held constant and is controlled by the pump setting. Under these conditions, the chromatogram will show the elution of A and B as a function of time after loading the sample (injection time). Many investigators change the variable elution time to the more useful parameter elution volume by taking the flow rate expressed in milliliters per minute and multiplying it by the reciprocal of the chart speed, expressed in minutes per centimeter, to give the new unit of milliliters per centimeter. This allows the length unit (cm) on the chart to be converted to volume. The elution volume is especially useful if it becomes necessary to change the flow rate from run to run or if chromatograms obtained in different laboratories under different flow rates are to be compared.

Figure 2.4 illustrates the usefulness of the volume unit. Figure 2.4A and B show chromatograms of the same sample obtained at different flow rates. Thus, whereas in Fig. 2.4A the sample was eluted at a flow rate of 2 mL/min, in Fig. 2.4B the flow rate was 1 mL/min. A superficial analysis of these data would suggest that the two peaks in panel A and the two

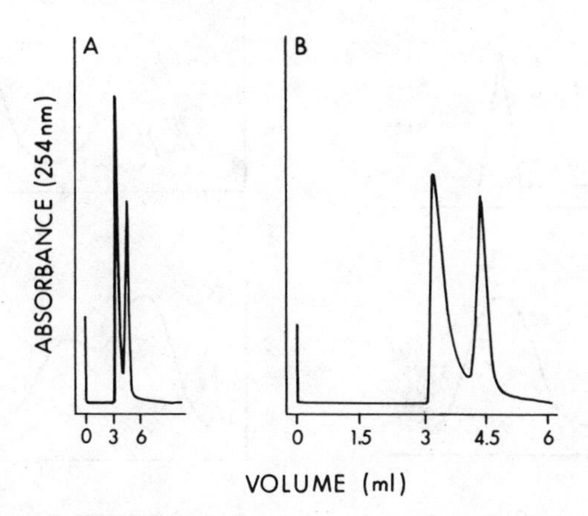

Figure 2.5 Data from Fig. 2.4 replotted as a function of elution volume. The volume was determined by the multiplication of the flow rate (mL/min) by the reciprocal of the chart speed (cm/min). Expressed in this manner, each unit of distance is converted to a unit of volume.

in B represented four different compounds. However, if these same data are expressed as a function of elution volume as shown in Fig. 2.5, then the peaks in panels A and B are easily seen to have similar retention volumes and thus, based on this criteria, are the same.

2.6 INTERPRETATION OF THE CHROMATOGRAM

In addition to showing that species such as A and B have been separated, chromatograms provide other information. For example, the shape of a curve provides information about the efficiency of the separation. With a system operating at high efficiency, peaks will be narrow and spikelike (Fig. 2.6A), while broad-based peaks suggest low efficiency (Fig. 2.6B). These results may be due in part to such factors as a lack of uniformity in either the size or homogeneity of the particles used in the stationary phase. Alternatively, broad peaks could indicate heterogeneity in the sample, as is often observed when the mobile phase is at a pH too near the p*K* of the molecules being separated.

The appearance of the peaks on chromatograms can also provide information about the quality of the resolution. Thus, if the compounds are well separated, the second peak will emerge only after the detector has

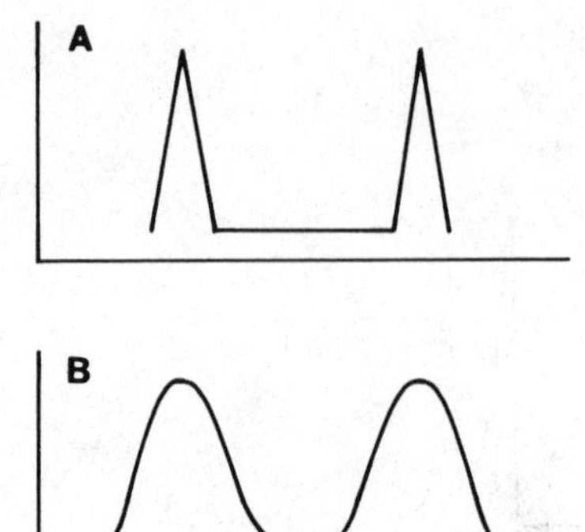

Figure 2.6 HPLC profiles of two components separated on two columns operating at different efficiencies. (A) A column operating at high efficiency; (B) a column operating at low efficiency.

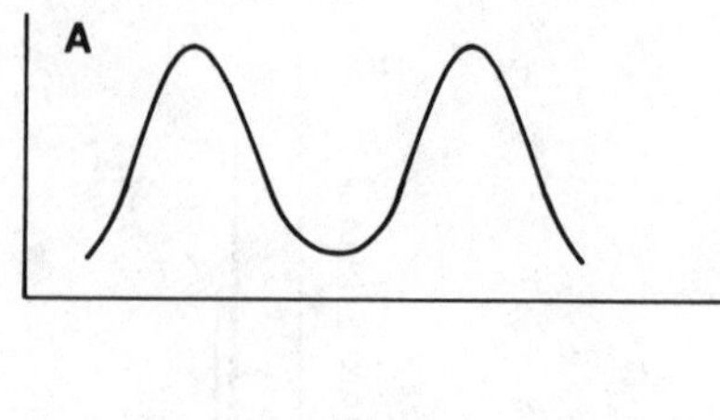

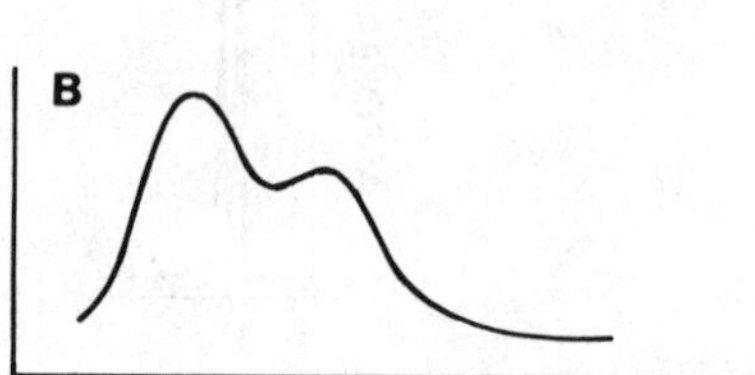

Figure 2.7 HPLC profiles of two compounds separated on column showing (A) good separation (resolution); (B) poor resolution.

completely returned to the baseline (Fig. 2.7A). Failure to achieve baseline separation (Fig. 2.7B) indicates poor resolution and suggests that something must be done to allow the second component to be retained longer—either slow its rate or increase the distance it must travel.

The resolution of any two components, therefore, is a ratio relating the distance between the apex of the peaks and the distance between their bases. With baseline separation, the bases of the peaks do not overlap. In the absence of baseline separation, however, the apex of each peak may be separate while the bases overlap. A mathematical expression can be written to describe this relationship by dividing the distance between the peaks, shown by the symbol delta (Δ) in Fig. 2.8, by one-half the sum of the width of the bases, giving a numerical value for resolution (Fig. 2.8).

The symmetry of each peak can provide information about the sample. Tailing (Fig. 2.9A) suggests some heterogeneity in the sample—either real or introduced by the chromatographic conditions. Flat-topped peaks (Fig 2.9B) suggest that the capacity of the column has been exceeded.

Of course, the magnitude of the signal from the detector can be used as a measure of the relative amount of each sample. While arbitrary units of

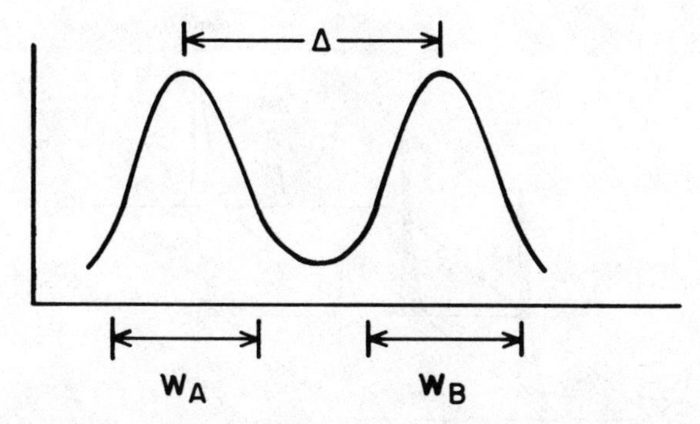

Figure 2.8 Representative hplc chromatogram to illustrate a method for calculation of resolution R. The separation of two components (labeled A and B) is shown. The width of each peak is shown by arrows and the symbol W, while the distance between peaks is shown by the symbol Δ. Resolution may be defined as

$$R = \frac{\Delta}{(^1/_2)(W_A + W_B)}$$

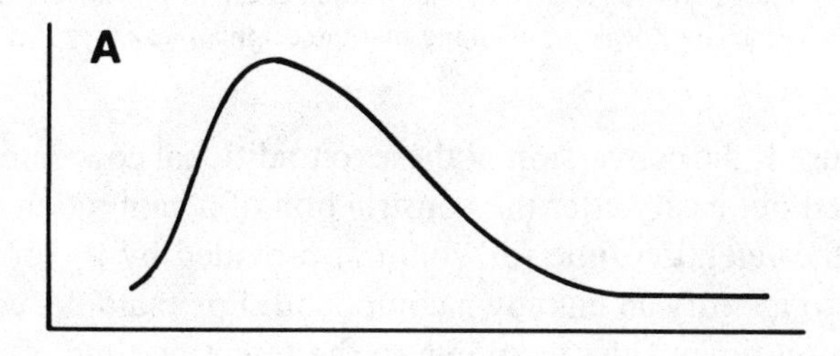

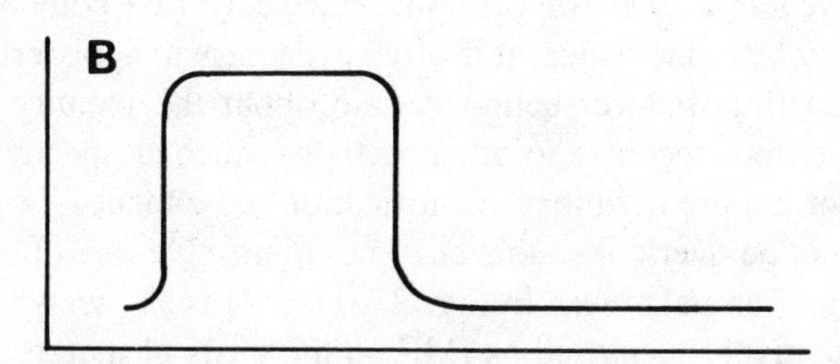

Figure 2.9 Hplc chromatograms of a single component showing (A) an example of "tailing" and (B) the profile obtained when a column is overloaded.

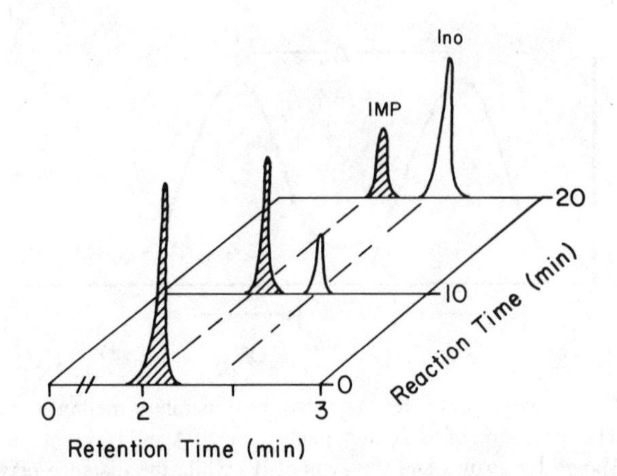

Figure 2.10 The use of enzymes to identify an unknown compound. In the illustration the compound was tentatively identified as IMP, based on its retention time of 2 min (chromatogram obtained at zero reaction time). The compound was incubated with a commercially available preparation of 5′-nucleotidase. Samples of the incubation mixture were removed and analyzed by hplc. The chromatograms illustrated, obtained at 10 and 20 min of reaction time, showed a reduction in the area of the IMP peak and an increase in the area of the inosine (Ino) peak, confirming that the original peak was IMP.

area can be used, the conversion of these to traditional concentration units can be carried out easily after the construction of a calibration curve.

Finally, the retention time (or volume) provided by a chromatogram can be used to identify an unknown compound. For example, comparison of the retention time of the unknown to the retention time of a series of standards (i.e., known compounds) is often sufficient to identify the unknown. However, a word of caution. Since any two compounds may coelute merely by coincidence, it is often necessary to apply criteria other than retention time before feeling certain about the identity of an unknown. Sometimes recourse to other methods, such as spectral analysis, is required for a more definitive identification to be made.

It should not be overlooked that enzymes themselves are often of use in identification of an unknown. In Fig. 2.10 a case is shown where a compound was tentively identified as IMP on the basis of its retention time. This conclusion was subjected to further testing using the enzyme 5′-nucleotidase, with the expectation that if the compound was IMP the enzyme would catalyze the removal of the phosphate and the formation of inosine. The chromatogram obtained following the addition of the enzyme and incubation for about 20 min is shown in Fig. 2.10. The chro-

Table 2.2 Selection of the Stationary Phase

Property	Separation
Size, shape	Gel filtration
Solubility, polarity	Reverse-phase
Charge, polarity	Ion-exchange

matogram now shows in addition to the IMP, which is reduced in amount, the appearance of a new peak with the correct elution time expected for inosine, the amount of which increases with incubation time (Fig 2.10). These data add credibility to the claim that the starting material was IMP.

2.7 THE SELECTION OF THE STATIONARY PHASE: SOME HELP FROM AN UNDERSTANDING OF THE PROCESS OF SEPARATION

While the selection of a stationary phase to be used in the analytical column may appear complex, the decision can be greatly simplified by considering the three basic methods of separation currently in use. These include (1) gel filtration or size-exclusion separation, (2) reverse-phase or hydrophobic separation, and (3) ion-exchange separation. In general, each type of separation uses a different kind of packing material, and since each type of separation exploits a different property of the molecules, the choice of packing really comes down to which property of the molecules would be most useful in achieving the separation.

For example, to use size-exclusion chromatography, the compounds to be separated must differ in size, shape, or both, while to use solubility or charge, the compounds must differ in polarity or net charge, respectively (Table 2.2). While deciding on a stationary phase using these as the only criteria clearly represents an oversimplification, and the reader is referred to specific works for more details, this approach can go a long way toward making the selection of the analytical column considerably easier.

2.7.1 Gel Filtration Chromatography

To understand gel filtration chromatography, imagine an analytical column packed with beads as shown in Fig. 2.11A. If a single bead were ex-

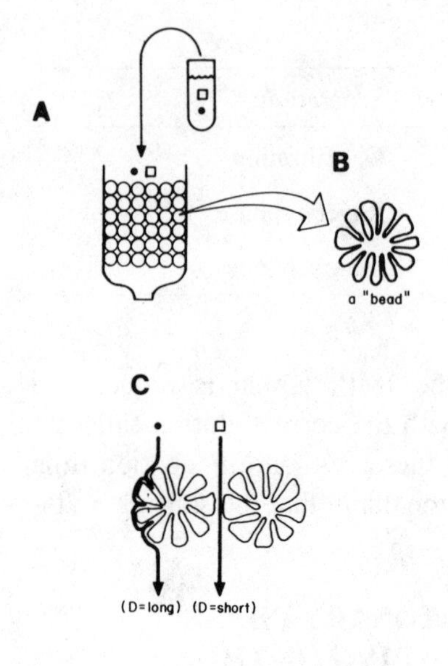

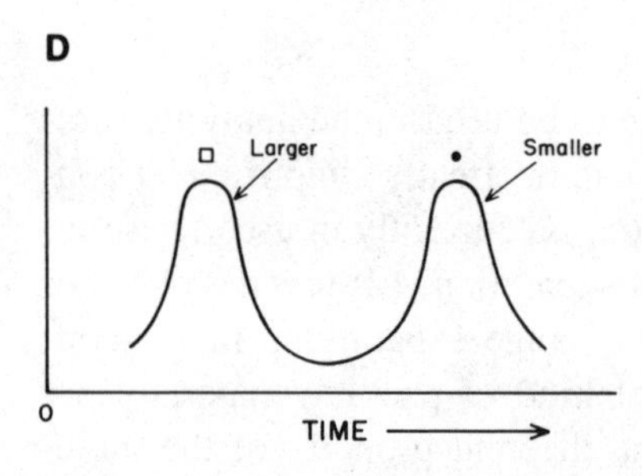

Figure 2.11 Gel filtration chromatography. In the illustration the sample contains two compounds that differ in size: (□) the larger compound; (●) the smaller. (A) The two compounds being loaded onto a column packed with beads, which, when viewed at the ultrastructural level, would appear as depicted in (B). The larger compound passes between the beads, while the smaller enters the crevices of each bead (C). (D) The chromatographic profile illustrates that the larger molecule will emerge from the column in less time than the smaller.

amined by scanning electron microscopy, a representation of its image might look like that shown in Fig. 2.11B. Each bead would be seen to have an irregular surface through and around which the mobile phase can enter and exit, in effect making the interior of the bead accessible to the mobile phase.

What about the sample? Imagine a sample to be composed of two types of compounds, the molecules of which differ in size as represented in Fig. 2.11. Now imagine that because of their differences the smaller molecules can follow the mobile phase as it meanders through and around the irregularities of the beads (Fig. 2.11), while the larger cannot. Thus, the larger molecules will be excluded from taking the longer path, and as a result of this exclusion the path, or distance *D*, followed by the larger molecules through the column will be short, and the larger or excluded molecules will exit the column first. The volume of solvent required for these molecules to emerge is spoken of as the *included volume* (Fig. 2.11D). The smaller molecule will follow a longer path and will emerge later. Because, in the ideal case, the molecules of neither compound will interact with beads, the difference in their time of emergence reflects the additional distance traversed by the smaller compound through the beads. Of course, altering the size of the irregularities of the beads will alter the size of the compounds excluded and therefore change the operating range of the analytical column.

2.7.2 Reverse-Phase Chromatography

As mentioned above, chromatography requires two phases: one solid and localized to the analytical column, the other mobile—the eluent or buffer that flows around and through the packing. As used by early workers, the packing was made of a material that was basically polar, while the mobile phases were nonpolar organic solvents. This arrangement of a polar stationary phase and a nonpolar mobile phase is, by virtue of tradition, referred to as *normal-phase* liquid chromatography. When the situation is reversed and the analytical columns are packed with a stationary phase that is nonpolar and are eluted with polar (aqueous) buffers, it is fortunate that this type of liquid chromatography was not referred to as "abnormal." Instead, since the phases have been reversed, this type of chromatography came to be known as *reverse-phase*. More recently the term "hydrophobic" has been suggested.

In describing its underlying mechanism of operation, it is convenient to focus again on the compounds within the sample and their movement through the analytical column. However, unlike gel filtration, where the order of elution of the compounds was determined by their path or the distance they traveled, in reverse-phase chromatography all the compounds in the sample travel the same path. In this case it is the rate with which they move through the column packing that determines the order of elution. Thus, a molecule that moves at a slower rate is said to be retained, and its time of elution is referred to as its *retention time*.

To understand the operation of the reverse phase, it might be useful to consider the illustration in Fig. 2.12, where a series of oil droplets (or

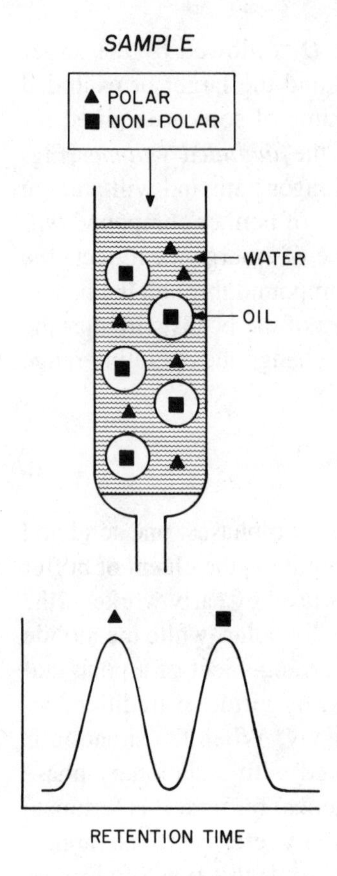

Figure 2.12 Reverse-phase hplc. In this illustration the sample is composed of two compounds, one polar (▲), the other nonpolar (■). The column packing (stationary phase) is illustrated as spheres and labeled "Oil" and the mobile phase as wavy lines labeled "Water". The polar molecules are shown remaining in the mobile phase (water), while the nonpolar molecules "enter" the stationary (oil) phase. Finally, the chromatographic profile illustrates that in this case the polar molecule will not be retained and will emerge with a shorter retention time than the nonpolar molecule.

beads) are suspended in a column filled with an aqueous buffer. The water represents the mobile phase, while the oil represents the stationary phase in the analytical column. Load the sample, in this case molecules of compounds A and B, onto the surface of the column, and the question becomes whether the compounds remain in the water (mobile phase) as they flow through the column, or whether they enter the beads of oil. Of course, if the compound remains in the water, its rate of progress through the column is effectively the flow rate of the mobile phase. There will be no interaction and this compound will not be retained. It will emerge soon after loading and will have a short retention time.

In contrast, by entering the oil, a compound leaves the mobile phase and interacts with the beads: its rate of passage through the system is in

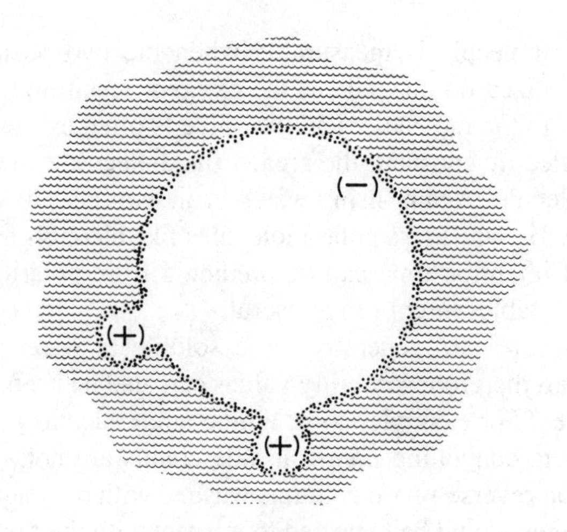

Figure 2.13 A representation of a water molecule to illustrate polarity. The position of each of the two positively charged protons is shown as (+). The position of the negative charge of the oxygen atom is shown as (−). The asymmetric distribution of the positive and negative charges produces the polarity.

effect slowed. The compound will be retained, and it will have a longer retention time than a compound that does not interact.

A great deal of time and effort has been spent in trying to predict the retention time of compounds in the reverse-phase system. While some rules have emerged and some generalizations made, to date the best approach remains a few trial runs.

The most useful parameters to consider when developing some feel for the operation of this type of chromatography are polarity and the related parameter solubility. Values of both parameters have been published for many of the compounds used in biological systems.

In what follows, I introduce the notion of polarity and, after differentiating it from the net charge of a molecule, use it to explain the retention time for some classes of compounds.

Polarity should not be confused with any net charge a molecule might have. For example, in some cases, highly polar molecules contain no net charge. Polarity is a result of an electrical asymmetry caused primarily by the electrons being distributed in an asymmetric fashion. A case in point is the water molecule, as shown in the representation in Fig. 2.13. The polarity of a molecule is measurable when the two positive hydrogen atoms are localized on one side of the oxygen, resulting in negative and positive sides to the molecule, and the value is often expressed as a function of its dielectric constant: the greater the dielectric constant, the more polar the molecule. Again, using water as an example, its dielectric constant is about 81, while less polar molecules like alcohols have lower values. Thus, if retention time can be predicted from polarity or dielectric constant, such tables might prove useful.

Polar molecules are generally more soluble in water than nonpolar molecules, and therefore solubility values can also be useful in predicting retention times. For example, some amino acids such as glycine and alanine and others containing nonpolar side chains are not very soluble in water; thus, on reverse-phase columns washed with only aqueous buffers, such compounds would be expected to interact with the nonpolar packing and be retained. Elution would be promoted by increasing the organic composition of the elution buffer.

Similarly, a comparison of the solubilities of some nucleobases in carbon tetrachloride and water will show that adenine is more soluble than guanine in organic solvents. Therefore, adenine will more likely enter the oil phase of the analytical column and be retained longer than guanine. Additional data obtained by measuring the distribution (or solubility) of a compound in either octanol or water shows that adenosine is about ten times more soluble than inosine in octanol. Again based on these findings, we might expect that adenosine will have a longer retention time than inosine, and in fact it does.

Similarly, consider the compounds adenosine and ATP. As is well known to most biologists, ATP has greater solubility than adenosine in aqueous buffers. This knowledge, therefore, can be of value in predicting the behavior of such compounds in the reverse-phase system. Figure 2.14 shows the elution sequence of ATP and adenosine on a reverse-phase column eluted with an aqueous buffer; ATP elutes significantly before adenosine. In fact, such a short retention time suggests that ATP has great difficulty interacting with the nonpolar stationary phase.

Adenosine, however, being less soluble (or more nonpolar) than ATP, will enter the stationary phase, and this is reflected in a longer retention time (Fig. 2.14). However, if a more nonpolar mobile phase were used to elute the column, the retention time for the adenosine would shorten, as it would remain longer in the mobile phase. Thus, if a mobile phase was be-

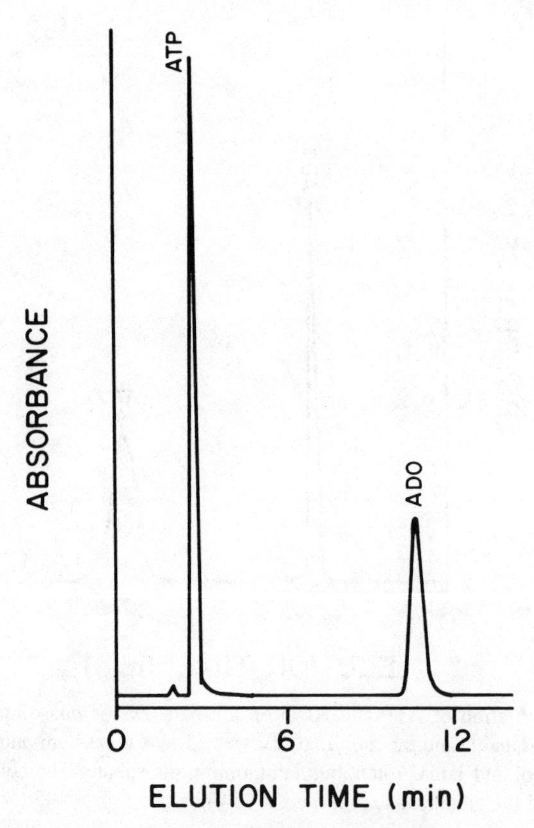

Figure 2.14 The separation of ATP and adenosine (ADO) by reverse-phase hplc. The prepacked column was C-18 (μBondapak), and the mobile phase was a 10 m*M* potassium phosphate buffer (pH 5.5) containing 20% methanol. The column was eluted isocratically and monitored at 254 nm. The flow rate was 2 mL/min.

ing used in which it was found that adenosine had as short a retention time as ATP, the separation of the two would be encouraged by reducing the amount of organic solvent in the mobile phase and causing the retention time of the adenosine to increase relative to that of ATP.

An interesting and useful variant of reverse-phase hplc is called *ion-paired reverse-phase hplc*. In such a system the analytical columns are packed with the same material, but a compound such as tetrabutylammonium is added to the mobile phase. The separation of ATP and adenosine on such a system is shown in Fig. 2.15. A comparison of this

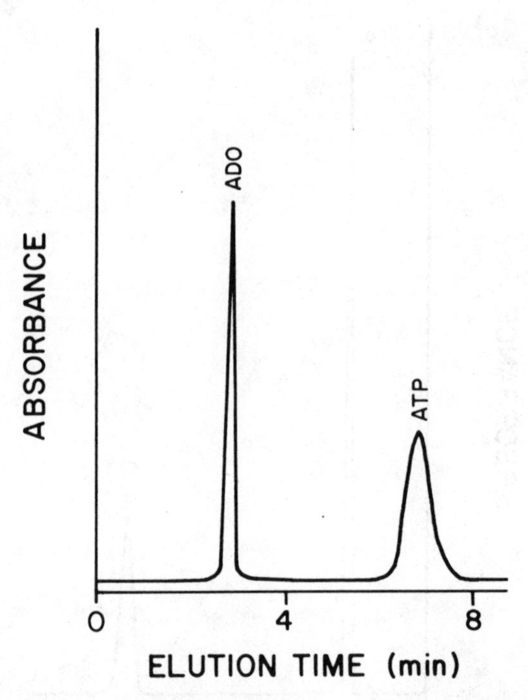

Figure 2.15 Separation of ATP and ADO on ion-pair reverse-phase hplc. The column was C-18 (μBondapak), and the mobile phase was 65 m*M* potassium phosphate (pH 3.7) with 5% methanol and 1 m*M* *n*-tetrabutylammonium phosphate. The column was eluted isocratically, and the eluent was monitored at 254 nm.

profile to that shown for the same compounds in Fig. 2.14 should immediately make apparent the change in the elution sequence. Whereas without ion pairing, the order is ATP followed by adenosine, with ion pairing the order is now adenosine followed by ATP.

An explanation for the difference in retention time can be developed if one imagines the tetrabutylammonium compound, which is positively charged, paired with the negatively charged ATP molecule. While this pairing will, in fact, reduce the net charge, the reduction in net charge will also reduce the polarity of the ATP molecule. Since the short retention time initially was a result of the polarity, any reduction in polarity would be expected to increase retention time. Thus, coming full circle, the effect of the tetrabutylammonium salt on retention times might be explained by its effect (reduction) on polarity.

In a series of experiments designed to explore further the role of polarity in affecting retention time in reverse-phase chromatography, we developed chemical procedures for the condensation of molecules of known polarity, expecting, for example, that joining two polar molecules should produce a relatively nonpolar molecule. In our first experiment we coupled the very polar nucleoside monophosphate AMP to lysine, an amino acid with a very polar side chain. The behavior of the two starting compounds in reverse-phase hplc is shown in Fig. 2.16A. Both have relatively short retention times consistent with their polar character. However, when the retention time of the conjugate was determined, it was found to be longer than that of either of the starting compounds (Fig. 2.16A). Thus, the combination of two polar compounds can produce a compound more non polar than either of the parent compounds.

Similar experiments were undertaken joining AMP to a dipeptide hippuryllysine. This particular dipeptide was used because a comparison of the retention time of lysine to that of hippuryllysine revealed that the addition of the hippuric acid to the lysine reduced its polarity. However, a determination of the retention times of the dipeptide and AMP on a reverse-phase column (Fig. 2.16B) reveals both to be polar. Nevertheless, their conjugate has a longer retention time than either of the starting materials. Note, however, that the decrease in polarity of this conjugate is very much less than what was observed following the summation of the AMP and lysine (Fig. 2.16A).

Finally, the AMP was coupled to the tetrapeptide tuftsin, which has the amino acid sequence Thr-Lys-Pro-Arg. Based on its extremely long retention time on a reverse-phase column, the tuftsin can be considered a nonpolar molecule, a conclusion supported by its rather low solubility in aqueous systems and the requirement for 40% methanol to elute it from the column (Fig. 2.16). When AMP is condensed onto the tuftsin, usually a single AMP per molecule of tuftsin, the polarity of the tuftsin is significantly decreased, as indicated by the decrease in the retention time of the conjugate. As shown in Fig. 2.16C, the conjugate has a retention time much closer to that of AMP than to that of tuftsin. This finding suggests that the addition of the polar AMP to the nonpolar tuftsin decreases the overall electrical asymmetry but does not eliminate it completely. Thus, while the combination of two polar molecules can produce a nonpolar molecule, so too can the combination of a polar with a nonpolar molecule produce a molecule more polar than its nonpolar parent.

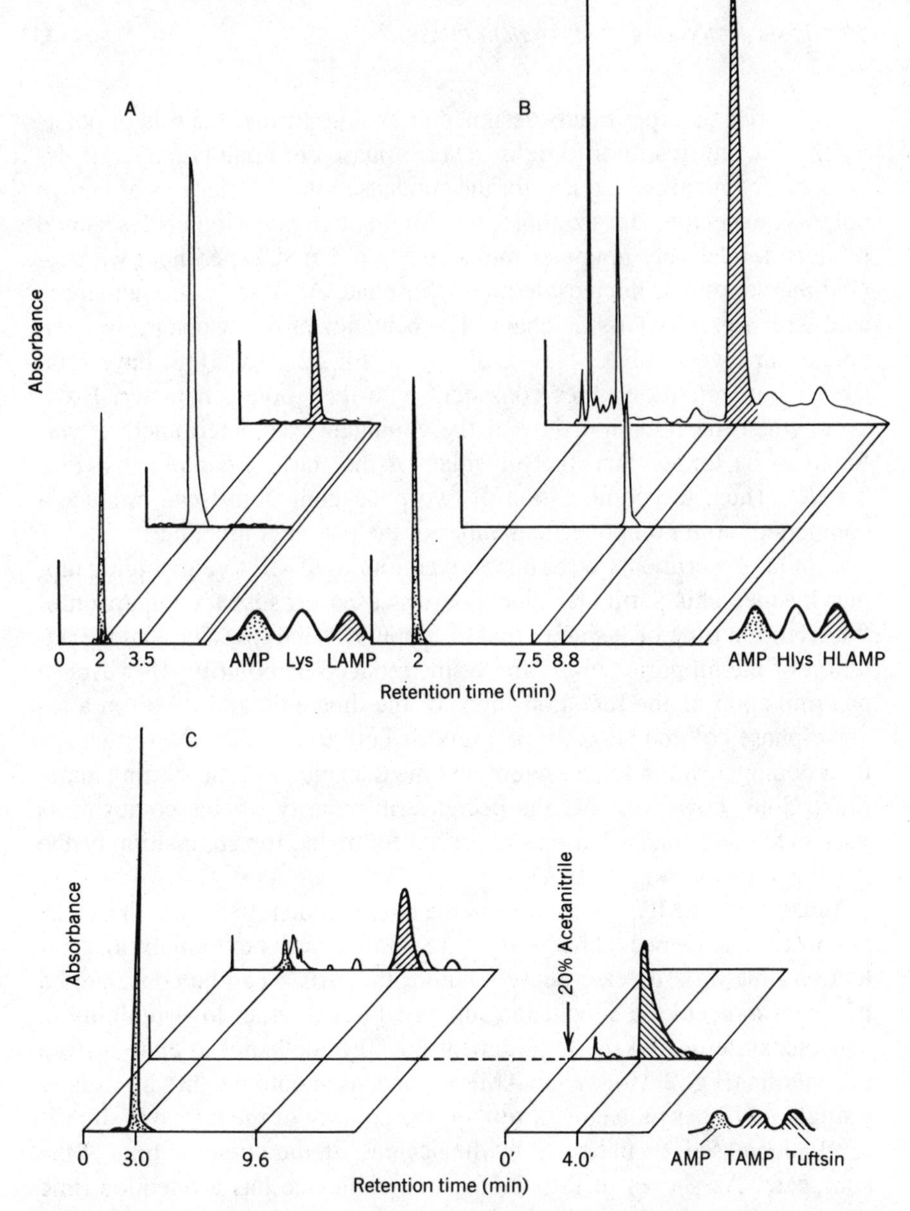

Figure 2.16 Effects of polarity on retention time. Hplc chromatography carried out on reverse-phase C-18 (μBondapak) column. (A) Chromatographic profiles of AMP, lysine, and the lysyl-AMP conjugate obtained using a mobile phase of 65 m*M* potassium phosphate (pH 3.6) and 2% acetonitrile. (B) Profiles of AMP, hippuryllysine, and the hippuryllysyl-AMP conjugate eluted with a mobile phase of 10 m*M* potassium acetate (pH 7.2) containing 1% acetonitrile. (C) Chromatograms obtained with AMP, tuftsin, and tuftsin-AMP. Compounds were eluted with a mobile phase of 65 m*M* potassium phosphate (pH 3.6) and 2% acetonitrile for AMP and tuftsin-AMP. Tuftsin was eluted by 20% acetonitrile. Detection was at 230 nm for lysine and 254 nm for all others.

It should be noted that reverse-phase hplc has been used to deduce polarity as, for example, in a study of cAMP and its analogs, as well as to predict partition coefficients and lipid solubility.

2.7.3 Ion-Exchange Chromatography

In ion-exchange chromatography, as in reverse-phase hplc, the rate at which a molecule moves through the analytical column and its interaction with the packing determine the order of elution of given compounds. In this case it is both the number and magnitude of the charge that influence interactions. The principles and method of operation in ion-exchange hplc are similar to those of the more conventional ion-exchange systems.

In general, the support (stationary) phase carries either a positive or negative charge. During equilibration of the column with the eluent, a counterion is introduced. The molecules to be separated must also be charged, and when the sample is loaded they bind to the fixed charges of the column packing and displace the counterion. Elution of the bound molecules is brought about by a second counterion, which is usually introduced as salt onto the packing by adding it to the elution buffer. The ability of the counterions (salts) to displace bound molcules relies on the difference in their affinities for the fixed charges of the stationary phase.

The interaction between the fixed charges of the stationary phase and the compounds adenosine, AMP, ADP, and ATP with zero, one, two, and three charges, respectively, is shown schematically in Fig. 2.17. In anionic-exchange chromatography, adenosine, with no charge, is not retained; it will be eluted in the absence of the addition of any salt. Thus, as represented in Fig. 2.17, adenosine (Ado) will have a short retention time.

Increasing concentrations of chloride will be required to displace and elute in series, AMP, ADP, and even higher concentrations of ATP (Fig. 2.17). Thus the elution order will be Ado, AMP, ADP, and ATP, with Ado having the shortest retention volume of the four (see Fig. 2.17). Such an elution order is consistent with the explanation that increasing the number of charges of a molecule increases its interaction with the stationary phase packing, thereby reducing its flow rate through the analytical column.

In considering the effectiveness of the different salt cations (or anions) in displacing or exchanging bound molecules requires a discussion of the

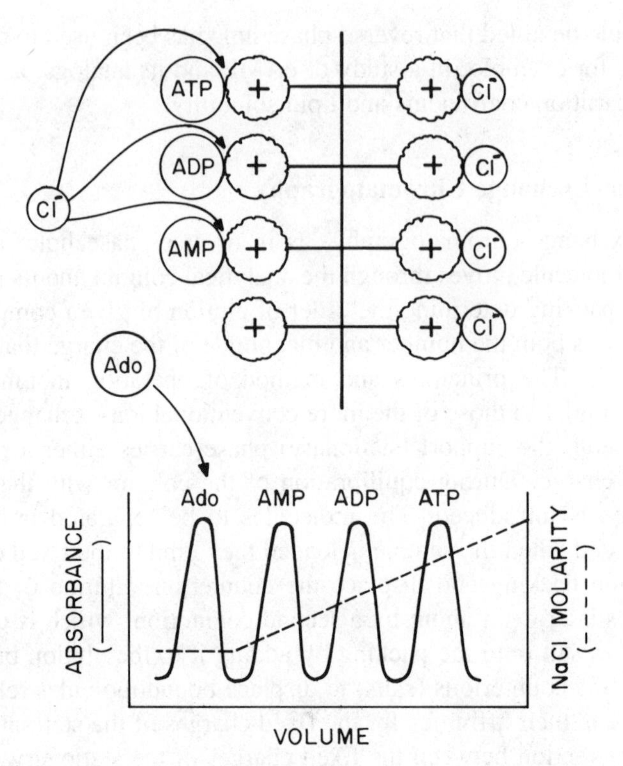

Figure 2.17 Representation of operation of ion-exchange chromatography. Top: Column functional groups, or fixed charges, are represented by serrated edge circles carrying a positive charge. They are shown "fixed" to a lattice. The compounds to be fractionated include ATP, ADP, AMP, and Ado. The first three are shown bound to the fixed charges, while the Ado is shown unbound. The introduction of the counterion, a chloride, is shown displacing the bound molecules from the fixed charges. Bottom: the order of elution as a function of the NaCl molarity, which is represented by the dashed diagonal line across the figure. The relative elution position (volume) of each of the four compounds is shown.

magnitude of the charge. Thus, molecules with one charge are not all equal when it comes to interacting with the fixed charges of the column packing. As a first approximation, the strength of the interaction may be considered in terms of the number of water molecules between or surrounding the salt ion; this is its hydrated radius.

For example, as represented in Fig. 2.18, cesium has only a few water molecules surrounding it compared to the lithium ion, which has many more (Fig. 2.18). One might think of these water molecules as a shield, with their elimination required for any interaction to take place between

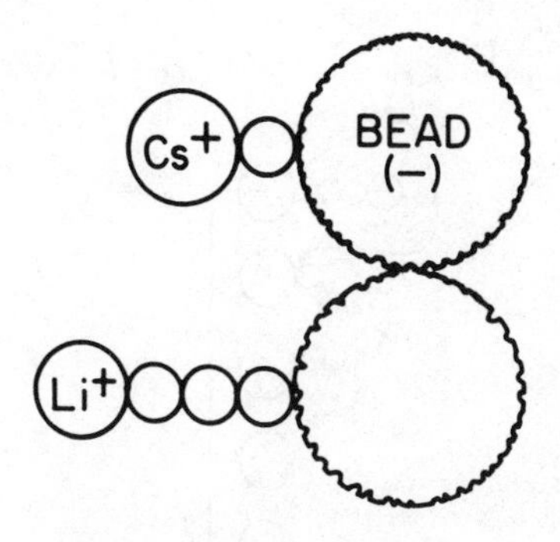

Figure 2.18 Effect of radius of hydration on distance between counterion and fixed charge of ion-exchange stationary phase. Cesium, with smaller radius of hydration, is shown with one water molecule (small circles) between it and fixed charge of the bead. Lithium is shown with three water molecules.

the ion and the packing. Clearly, it takes less energy to eliminate one molecule, and therefore it is not surprising that cesium is more effective than lithium at displacing molecules such as ATP bound to ion exchangers. This effectiveness is seen operationally in terms of the concentration of the counter ion required to elute the bound sample. It is also not surprising that affinities can be affected by modifications that alter the water content of the system, for example, by increasing salt concentrations, temperature, or the organic solvent content of the mobile phase. However, on an ion-exchange hplc column, the AMP is eluted at a lower salt concentration than cAMP, as illustrated in Figure 2.19. This differ-

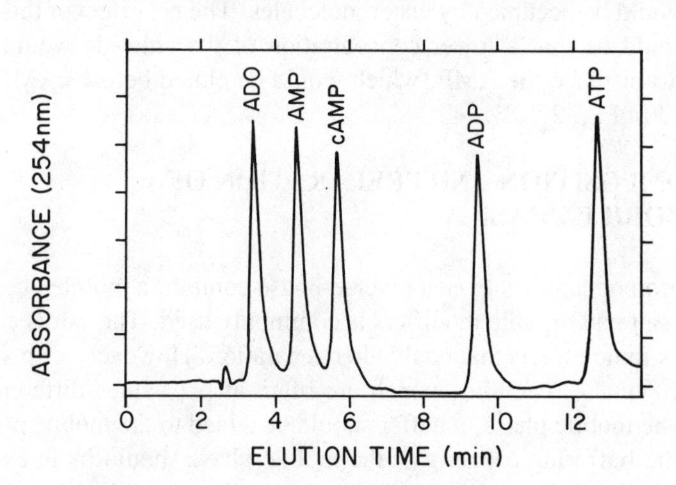

Figure 2.19 Separation of several nucleosides on ion-exchange hplc. HPLC was carried out on an ion-exchange column (AX-100) eluted isocratically with a mobile phase of 0.1 *M* sodium phosphate buffer (pH 7.3) containing 0.8 *M* sodium acetate. The column was monitored at 254 nm. A standard solution containing approximately 2 nmol each of adenosine (ADO), AMP, cAMP, ADP, and ATP was loaded onto the column.

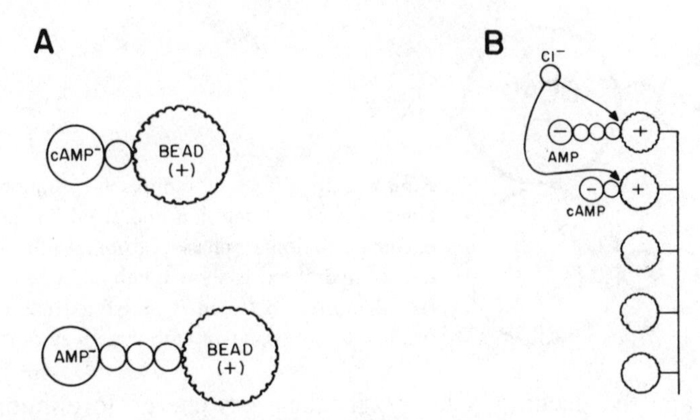

Figure 2.20 Representation of the interaction between the fixed charges of the ion-exchange beads and the cAMP and AMP molecules.

ence in affinity might be explained by the model described above in which relative affinity is a function of the distance between the mobile ion and the fixed ion. In this model, AMP, with less affinity, would have a greater distance between it and the fixed charge on the bead than cAMP would have. In both cases, as illustrated schematically in Fig. 2.20A, the space would be occupied by water molecules. The net effect of this difference would be that a lower concentration of the chloride would be required to displace the AMP, which would be eluted before cAMP, as illustrated in Fig. 2.20B.

2.8 COMPOSITION AND PREPARATION OF THE MOBILE PHASE

For elution of samples from a reverse-phase column, a mobile phase containing salt and organic modifiers is commonly used. The salt is added to suppress ionic effects that could alter separation. However, with samples from enzymatic reactions, which are often at pH values different from that of the mobile phase, a buffer should be added to the mobile phase, as well. The buffering capacity of the mobile phase should be in excess of that in the incubation mixture. This excess will ensure that when the sample is injected its pH will equilibrate to that of the mobile phase. These buffers should be made with distilled, deionized water that has been degassed to remove any trapped air. Degassing prevents bubble formation, which otherwise occurs in the pump head, particularly at high pressures. These bubbles are one of the major causes of variations in pump pressure, which, in turn, can produce artifacts in the chromatographic profile. Degassing, accomplished by vacuum aspiration, should be carried out with constant stirring until only a few bubbles are formed. Usually about 20 min/L is adequate.

Phosphate in concentration ranges of 10–100 m*M* can be used not only for buffering but also for ion supression. However, some thought should be given to any later uses for the compounds purified by the hplc. For example, if after purification the component will be examined by phosphorus NMR, the use of a mobile phase containing phosphate should obviously be avoided. In addition, the use of phosphate in ion-exchange mobile phases can lead to high background absorbance in the UV range as a consequence of impurities in the phosphate buffers. Although methods have been described to reduce this background (see General References), where possible the phosphate should be eliminated. The pH can always be adjusted with KOH. With such bases, however, the choice of cation, for example, K or Na, should be made with the composition of the sample buffer in mind. For example, if the buffer contains sodium dodecyl sulfate (SDS), potassium, which will precipitate it, should be avoided. Also, since halides such as chloride can cause corrosion of the stainless steel tubing, they should be avoided. Finally, if the purified components are to be concentrated by an evaporation procedure, a buffer with volatile components should be chosen.

In order to reduce retention time of nucleotides on reverse-phase columns, organic modifiers such as methanol or acetonitrile may be added to the buffer. Acetonitrile is usually more effective than methanol in the sense that less is required to elute a given nucleotide with a specific retention time.

Once the buffer (mobile phase) of the appropriate composition has been made and the pH adjusted, it should be filtered through a 0.45-μm filter to remove particles that may clog the column head. Following filtration, the buffer may be used for about 2–3 days if stored at room temperature, although the pH should be checked and precautions taken to keep the organic modifer from evaporating during storage.

At the conclusion of each work day it is advisable to wash the salt buffer thoroughly from the both the pump heads and the column. A 0.02% sodium azide solution prepared with degassed water should be used to wash the system free of salts. About 15 min of washing time is adequate with a flow rate of 2 mL/min. The sodium azide is used to control bacterial growth. A reverse-phase column should then be washed with and stored in a methanol–water solution (80:20).

It should be noted that the washing procedure will remove material from the guard column and the top of the analytical column and that when this material passes through the detector it often causes changes in optical density. Therefore, the recorder should be left on during the washing to ensure the complete removal of this debris. The recorder will return to a baseline reading when the washing is complete. Note also that the viscosity of water and methanol differ. When shifting from one to the other this change in viscosity of the solutions will produce an increase in back pressure which could be significant in magnitude but is not worrisome. Also note that a sodium azide solution moving through a detector set at 254 nm will usually produce an increase in optical density. Again these are normal changes and should not be of any concern.

2.9 COLUMN MAINTENANCE

When reverse-phase columns are used for the analysis of enzymatic reactions, many of the components of the reaction may become bound to the packing material. As a result, the debris may alter retention time, chromatographic profiles, or both of subsequently injected molecules. Types of column malfunction include (1) peak splitting or the appearance of a shoulder; (2) loss of baseline resolution, broadening of peaks, particularly at their base, or both, and (3) an increase in back pressure. To some extent, all these symptoms may be traced to material that adhered to the column and is not removed during the methanol wash.

If divalent metals are suspected as the cause of the peak splitting, washing the column with 100 mL of 10 m*M* EDTA in 10 m*M* phosphate at pH 5.5 may help eliminate the problem by removing the metal.

Many components that bind can be eluted by changing the pH of the mobile phase. Thus, a wash consisting of 200 mL total volume, at 2 mL/min, of 100 m*M* phosphate solution ranging in pH from 2 to 8 is frequently useful. Finally, if the increase in back pressure is suspected to be a result of contamination from bound protein, washing the column with at least 100 mL of 6 *M* urea in 20 m*M* phosphate (pH 7.8) may eliminate the problem. Again, the return of back pressure to normal values can be taken as a sign of success of any one of these steps.

It should be noted that we have found urea to be a poor wash solution on columns packed with reverse-phase packings. Dimethyl sulfoxide (DMSO) has been found to be useful in some cases.

In addition, a gradient progressing from 100% methanol through a series of less polar, more organic solvents such as carbon tetrachloride will serve to remove other reversibly bound contaminants. A reverse gradient should be used to reequilibrate the column to standard conditions.

Note that following any of the maintenance procedures listed above, reequilibration of the column to the original mobile phase will be required and will probably involve additional time that might not be necessary for the normal routine change from water to the methanol solution used to prepare the column for overnight storage.

As an illustration, consider the problem of the contamination of a reverse-phase column with a very sticky dextran sulfate material that had been added as an activator for an enzyme reaction. The compounds AMP, ADP, and ATP were being separated using a mobile phase containing phosphate buffer, acetonitrile, and tetrabutylammonium ion. The separation usually obtained is shown in Fig. 2.21A. However, in the presence of dextran sulfate the separation was less than adequate, as shown in Fig. 2.21B. To regenerate the column, it was washed first with 6 *M* urea. Analysis of the sample produced is shown in Fig. 2.21C. Some improvement in the separation is evident. Next the column was washed with toluene. The result, seen in Fig. 2.21D, shows that this step completely restored the separation capabilities of the packing. Additional details may be obtained by consulting the General References at the end of this chapter.

2.10 MONITORING COLUMN PERFORMANCE

In general, new columns should be calibrated in the laboratory in which they will be used. For this purpose a standard mobile phase and a standard series of compounds should be available in the laboratory. The resolution obtained under the standard conditions of the laboratory at the start of the useful life of the column should be recorded together with the date of the analysis. All should be part of the record for that column. At the first sign

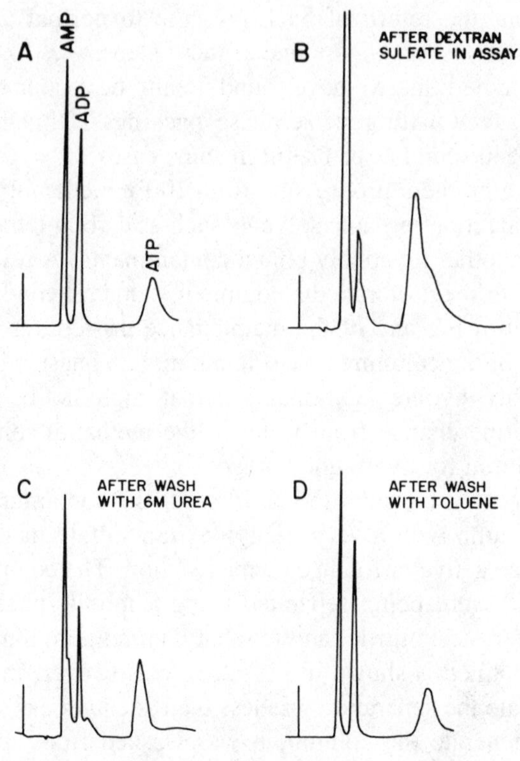

Figure 2.21 The effect of washing procedures on removal of debris as measured by separation of nucleosides. Separations were carried out on a reverse-phase (C-18) column with a mobile phase of 65 m*M* potassium phosphate (pH 3.6), 2% acetonitrile, and 1 m*M* tetra-*n*-butylammonium phosphate. The flow rate was 2 mL/min; the column was eluted isocratically and monitored at 254 nm. In this illustration the column was clogged with dextran sulfate. (A) Separation routinely achieved with AMP, ADP, and ATP; (B) separation observed after clogging the column with 10 m*M* dextran sulfate; (C) the separation observed after washing the column with 6 *M* urea; (D) the separation obtained after washing the column with toluene.

of problems with this column, its performance should be checked against these records, using the same mobile phase and standards.

2.11 SUMMARY AND CONCLUSIONS

Chromatography involves the separation of classes or groups of molecules. Two phases are usually required: one is stationary or solid, and the other mobile—eluent or buffer.

If the solid phase is in the form of particles or beads, it is usually packed into a tube or column and the buffer or mobile phase is pulled through the packing by gravity or forced through with a pump.

The time required for a given compound to emerge from the column is a function of the packing material and the interactions between the compound and the packing. Transition time is affected by the distance the compound travels or the rate at which it travels a fixed distance.

Other things being equal, resolution is enhanced by using smaller particles of the packing material. However, smaller particles result in tighter packing, which in turn requires higher pressures to push through the solvents. The combination of small particles with better pumps led to improved performance and the method called high performance liquid chromatography.

The basic equipment required for an hplc system include a solvent reservoir, a pump, an injector, an analytical column, a detector, and a recorder. The analysis of the sample is displayed as a chromatogram, with detector deflection presented usually as a function of time after loading the sample. By virtue of the shape of the curves, the distance between them, and their area it is possible to determine if the volume of the sample is too large, the number of different compounds that are present and the amounts of each of the compounds present in the sample. From an understanding of the process of separation, it is possible to select the appropriate stationary phase. Separation by gel filtration requires compounds of different sizes and shapes, while reverse-phase hplc will separate molecules that have different polarities. In contrast, ion-exchange hplc separates molecules with different charges.

The mobile phase contains salts and organic modifiers. A buffer is also required with enzymatic reactions to ensure a constant pH during the separation step.

Column maintenance will often require washing the column with chelators, denaturants, organics, or salt solutions of high concentration, all designed to remove debris bound to the column that is not removed by routine washing.

GENERAL REFERENCES

General References for HPLC

Brown, P. R., *High Pressure Liquid Chromatography: Biochemical and Biomedical Applications*, Academic, New York, 1973.

Hearn, M. T. W., ed., *Ion-Pair Chromatography: Theory and Biological and Pharmaceutical Applications*, Dekker, New York, 1985.

Henschen, A., Hupe, K., Lottspeich, F., and Voelter, W., *High Performance Liquid Chromatography in Biochemistry*, VCH, Deerfield Beach, FL, 1985.

Krstulovic, A. M., and Brown, P. R., *Reversed-Phase High-Performance Liquid Chromatography: Theory, Practice and Biomedical Applications*, Wiley-Interscience, New York, 1982.

Regnier, F. E., *Science* **222**:245 (1983).

Snyder, L. R., and Kirkland, J. J., *Introduction to Modern Liquid Chromatography*, Wiley, New York, 1979.

Polarity and Solubility

Cohn, E. J., and Edsall, J. T., *Proteins, Amino Acids and Peptides*, Hafner, New York, 1939.

Cullis, P. M., and Wolfenden, R., *Biochem.* **20**:3024 (1981).

Greenstein, J. P., and Winitz, M., *Chemistry of the Amino Acids*, Vol. 1, Wiley, New York, 1961.

Hansch, C., and Leo, A. J., *Substituent Constants for Correlation Analysis in Chemistry and Biology*, Wiley, New York, 1979.

Kolassa, N., Pfleger, K., and Rummel, W., *Eur. J. Pharm.* **9**:265 (1970).

Nahum, A., and Horvath, C., *J. Chromatog.* **192**:315 (1980).

Plaut, G. W. E., Kuby, S. A., and Lardy, H. A., *J. Biol. Chem.* **184**:243 (1950).

Reverse-Phase Chromatography

Hacky, J. E., and Young, A. M., *J. Liq. Chromatog.* **7**:675 (1984).

Hammers, W. E., Meurs, G. J., and DeLigny, C. L., *J. Chromatog.* **247**:1 (1982).

Hancock, W. S., ed., *Handbook of HPLC for Separation of Amino Acids, Peptides and Proteins*, Vols. I and II, CRC Press, Boca Raton, FL, 1984.

Krstulovic, A. M., and Brown, P. R., *Reversed-Phase High-Performance Liquid Chromatography: Theory, Practice and Biomedical Applications*, Wiley-Interscience, New York, 1982.

Perrone, P., and Brown, P. R., Ion-pair chromatography of nucleic acid derivatives, in M.T.W. Hearn, ed., *Ion-Pair Chromatography*, Dekker, New York, 1985.

Rossomando, E. F., and Hadjimichael, J., *Int. J. Biochem.*, **18**:481 (1986).

Care and Maintenance of Columns

Runser, D. J., *Maintaining and Trouble Shooting HPLC Systems.*, Wiley, New York, 1981.

Detectors

Henderson, R. J., Jr., and Griffin, C. A., *J. Chromotog*. **298**:231 (1984).

Ion-Exchange Chromatography

Jahngen, J. H., and Rossomando, E. F., *Anal. Biochem*. **130**:406 (1983).

Regnier, F., High-performance ion-exchange chromatography, W. B., Jakoby, ed., in *Methods in Enzymology*, Vol. 104, 170 Academic, Orlando, FL, 1984.

Preparation of Mobile Phase

Karkas, J. D., Germershauser, J., and Liou, R., J. *Chromatography* **214**:267 (1981).

Plunkett, W., Hug, V., Keating, M. J., and Chubb, S., *Cancer Res*. **40**:588 (1980).

Pumps: Operation and Troubleshooting

Dolan, J. W., and Berry, V. V., *LC Mag*. **1**:470 (1983).

Dolan, J. W., and Berry, V. V., *LC Mag*. **2**:210 (1984).

Chapter Three

Strategy for Design of an HPLC System for Assay of Enzyme Activity

Overview

In this chapter, a strategy will be presented for the design of an hplc assay system. Section 3.1, "Setting up the Assay," will focus on the enzymatic reaction and the steps leading to the development of the assay. These steps are previewed in Table 3.1. Section 3.2, "The Use of HPLC to Establish Optimal Conditions for the Enzymatic Reaction," will discuss the procedure for monitoring the activity of an enzyme with the hplc method and the use of the hplc assay method to determine the parameters required for obtaining optimal activity.

3.1 SETTING UP THE ASSAY

3.1.1 Analysis of the Primary Reaction

The design of an hplc assay system for an enzymatic activity begins with a complete analysis of the primary reaction—the reaction catalyzed by the enzyme under study. To begin this analysis, indicate all substrates, products, and cofactors of the reaction. If metals are required for catalysis, include them. In the case of the metals, however, it is useful to note

Table 3.1 Steps in Design of HPLC Assay for Enzymatic Reaction

1. Analyze the primary reaction.
2. Analyze all secondary reactions.
3. Select the mode of hplc (size-exclusion, ion-exchange, reverse-phase) that will allow for separation of substrates from products.
4. Make initial selection of mobile phase (pH, buffer, salt concentration) and method of delivery (isocratic or gradient elution).
5. Select appropriate detector. Will it be necessary to collect fractions?

whether they are an integral part of the substrate, for example, when the complex MgATP is the substrate, or whether they are required for some other function, such as activation of the enzyme. It is also useful to indicate the pH of the reaction as well as the type and concentration of the buffer to be used. The goal of this analysis is to list all the components present in the reaction mixture before the start of the reaction.

To illustrate this approach, consider the assay of a pyrophosphohydrolase, an enzyme that catalyzes the reaction

$$\text{MgATP} \rightarrow \text{AMP} + \text{PP}_i \quad (1)$$

MgATP is the substrate, and AMP and pyrophosphate (PP_i) are the products. Since this activity is usually assayed at a pH of 7.5 using a Tris-HCl buffer system, the reaction tube will contain ATP, Mg, and Tris-HCl as illustrated in Fig. 3.1.

3.1.2 Analysis of Secondary Reactions

The hplc method can be used to advantage in the assay of activities in crude extracts or in preparations only partially purified. Such samples will usually contain activities other than the one under study that will react with the substrate of the primary reaction or with its product, or with both. These other activities are referred to as secondary reactions. What are secondary reactions? They are reactions catalyzed by enzymatic activ-

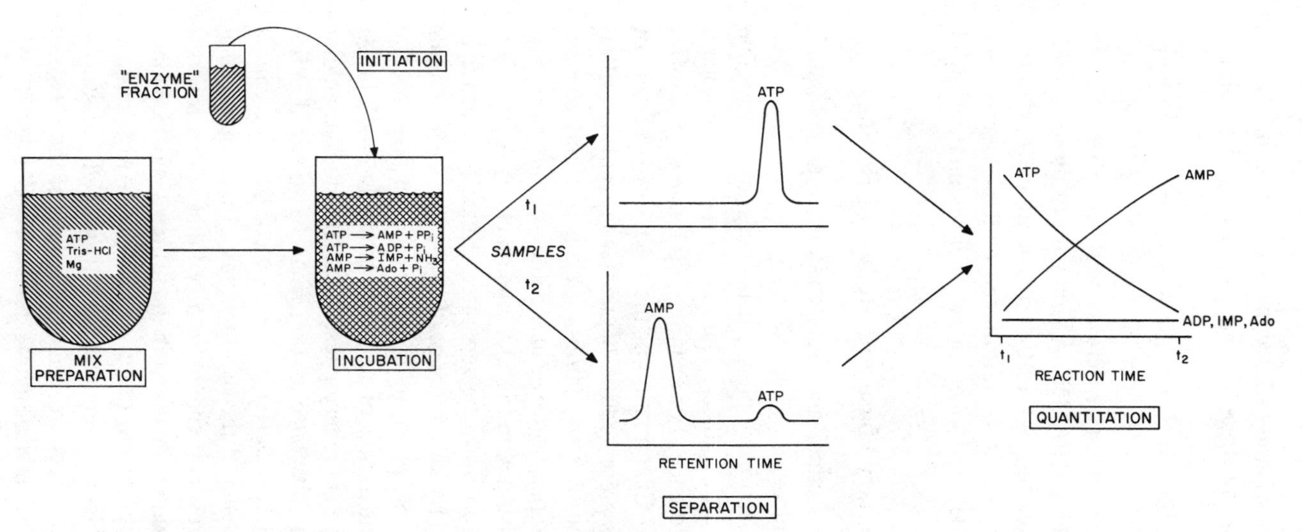

Figure 3.1 Overview of strategy for design of an hplc method to determine enzymatic activity. The reaction tube contains a mix preparation to measure the activity of an ATP pyrophosphohydrolase which catalyzes the formation of AMP and PP_i from ATP. The mix includes the substrate, ATP; the buffer, Tris-HCl; and magnesium, a metal cofactor. The addition of a sample from the "enzyme" fraction initiates both the primary reaction and several secondary reactions. Samples of the incubation mixture are withdrawn at intervals (t_1 and t_2), and the reaction is terminated by injection of the samples onto the hplc column. A representative analysis of each of the samples is shown. The amount of each component can be calculated from the area of each peak and is graphed as a function of reaction time.

ities other than the activity under study. A secondary reaction may use as substrate either the original substrate or the product of the primary reaction.

For example, AMP, the product of the primary reaction, (1), may undergo secondary reactions to form adenosine and phosphate or IMP and ammonia. In addition, other secondary reactions could involve ATP; an example is the degradation of ATP to ADP. A summary of these secondary reactions is also given in Fig. 3.1 in the step marked Incubation. While these secondary reactions can be eliminated or their significance minimized, they should not be overlooked in the analysis and design of the assay system.

3.1.3 Selection of the Stationary Phase and Method of Elution

With the list of reactants, cofactors, and reaction conditions compiled, the stationary phase can be selected.

To select the appropriate stationary phase, it is necessary to examine the reactants to determine how they differ. The selection of the stationary phase should exploit this difference. For example, do the reactants differ in size, charge, or solubility? Examination of the modes of operation of the stationary phase materials presented in Chapter Two reveals gel filtration to be ideal for separation by size, while ion-exchange and reverse-phase hplc are suitable for separation by charge and solubility, respectively. With the reactions illustrated in Fig. 3.1, the reactants ATP, ADP, and AMP, and adenosine (Ado) must be separated. Since they differ in charge, they can be separated by ion exchange. But they differ in solubility also, and therefore a reverse-phase hplc can be chosen instead. The choice is based on the consideration of another parameter, the elution procedure. With ion-exchange stationary phase, gradient elution, an elution procedure that varies in salt concentration, may be required. With reverse-phase columns, an elution buffer of constant composition can be used. Since the latter is the easier of the two methods to use, the reverse-phase hplc was chosen. However, it should be noted that with a reverse-phase hplc system, the compounds involved in the primary and secondary reactions shown above emerge in the order ATP, ADP, AMP, and, as shown in Fig. 3.2A, it is difficult to separate ATP from ADP. Since one of the goals of the separation procedure is the resolution of these two, it was decided to use ion-paired reverse-phase hplc (see Chapter Two). With this change, the ADP is eluted before ATP, as illustrated in Fig

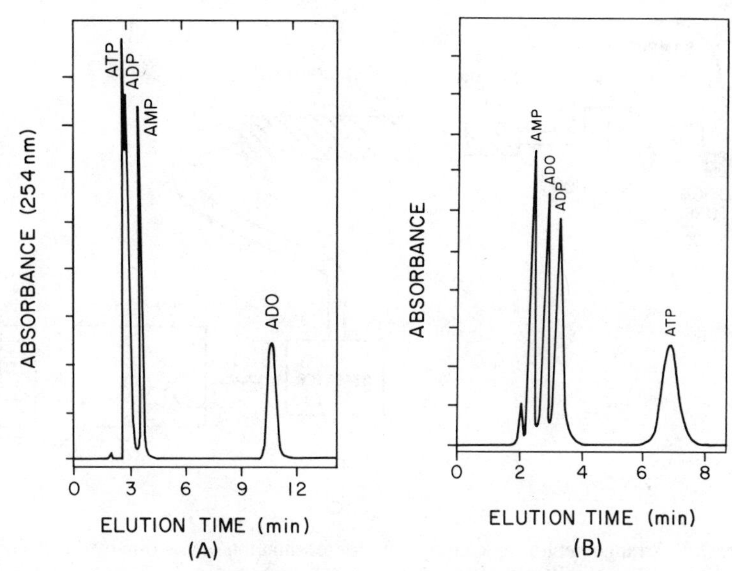

Figure 3.2 Analysis of ADO, AMP, ADP, and ATP by hplc. Separations were carried out on a reverse-phase C-18 column. (A) The mobile phase was composed of 10 m*M* potassium phosphate (pH 5.5) and 20% methanol as the mobile phase; (B) the mobile phase contained 65 m*M* potassium phosphate (pH 3.7), 5% methanol, and 1 m*M* *n*-tetrabutylammonium phosphate. In both A and B, the columns were eluted isocratically and the detection was at 254 nm.

3.2B. Therefore, when making the choice of the stationary phase, it is wise to consider the method of elution as well.

3.1.4 Modification of Reaction Conditions for the HPLC Assay Method

The reaction conditions used with other methods may have to be modified when the hplc assay procedure is to be used. It may be necessary to change the concentration of the various components of the system being examined, such as metals, hydrogen ions, or enzyme, particularly if the samples for analysis are taken directly from the incubation mixture and injected using equipment arranged as in Fig. 3.3. Note that when a sample is removed directly from the incubation mixture and is injected onto the hplc column for analysis, it brings with it everything present in the reaction mixture, including excess protein and metals, two components that

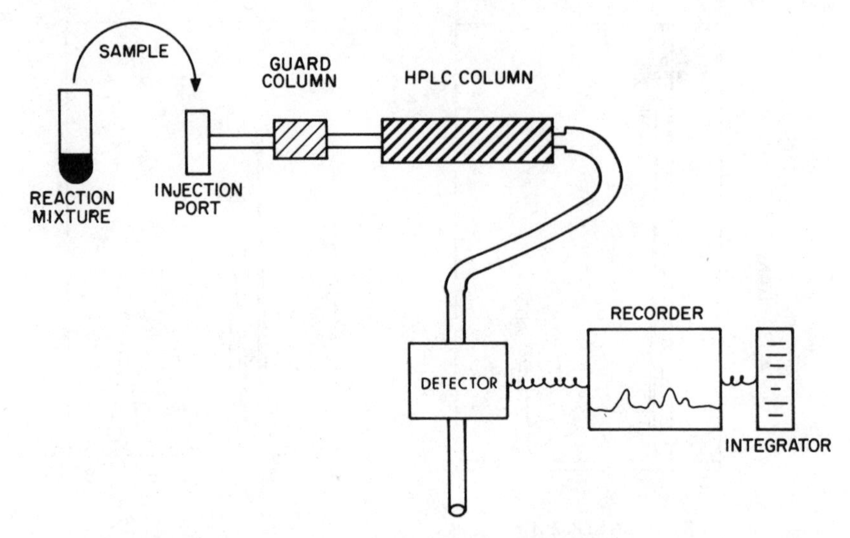

Figure 3.3 Arrangement of hplc equipment for termination of reaction by direct injection of sample. A sample is removed from the reaction mixture and is transferred directly to the injection port for introduction onto the column. The hplc column is protected by a guard column, which removes debris. The eluent flows through the detector, from which a signal is displayed on a recorder. The area of each peak is electronically integrated.

can clog the column or alter the performance of the column packing. Most, if not all, of these problems can be solved by terminating the reaction prior to injection. The direct analysis of samples may require reducing the metal concentration, working at lower enzyme concentrations, or even compromising with pH values to balance reaction conditions with hplc assay conditions.

3.1.5 Understanding and Dealing With Secondary Reactions

The importance of understanding secondary reactions cannot be overemphasized. This knowledge is invaluable to the interpretation of chromatographic profiles of enzymatic reactions. "Beware of secondary reactions" is a rule that should always be kept in the forefront. This rule should be remembered even when performing hplc analysis of purified enzymes obtained from commercial sources. Consider the case in which we studied a commercially available preparation of alkaline phosphatase. We decided to study the activity of this enzyme with AMP as the substrate. A reverse-phase column was used to separate the substrate from one of the expected products, adenosine. The addition of enzyme initiated the reaction, and samples were taken at intervals and analyzed by hplc. Several of the chromatograms that were obtained are shown in Fig. 3.4.

These chromatograms were studied with the expectation that the enzyme was pure. Therefore, we thought the chromatogram would show only the substrate AMP, and, as illustrated in Fig. 3.4A, this was the case. Chromatograms obtained late in the incubation were expected to show the reaction product adenosine (Ado). However, as illustrated in Fig. 3.4B and C, later chromatograms showed three peaks. Two peaks were easily identified—one was AMP, and the other adenosine. The third peak was ignored, since its area appeared to be insignificant. However, when we measured the amount of adenosine recovered (Fig. 3.4C), it did not equal the amount of AMP lost. The possibility that more of the reaction product Ado had remained on the column could be ruled out by using different mobile phases to elute all bound material. The chromatograms accounted for all the products. Since we had not expected any side reactions, we quantitated the yield of reaction products on the basis of area, assuming the presence of only adenosine-containing compounds. The formation of inosine, with a 50% reduction in extinction coefficient, could account for the apparent lack of recovery. Therefore, we considered the presence of secondary reactions. Either AMP had been converted to IMP, or adenosine was converted to inosine (Ino). Comparing the retention time of the third peak to authentic standards, we ruled out IMP as a product, and the identity of peak 3 was established as inosine. This led us to conclude that the commercial preparation of alkaline phosphatase was contaminated with a second activity, adenosine deaminase. To follow up the observation, a second reaction mixture was prepared that was similar in composition to the one used previously but to which we added an inhibitor of adenosine deaminase. After the start of the reaction, samples were removed and analyzed, and the chromatograms obtained (shown in Fig. 3.4D–F) illustrated the loss of AMP and quantitative recovery of the adenosine.

How can secondary reactions be handled? Some procedures are presented in Table 3.2. These include purifying the activity of the primary reaction to homogeneity. However, this may not always be possible, since activities must be assayed in crude extracts. Therefore some other solution must be found. The use of analogs is one such solution. For ex-

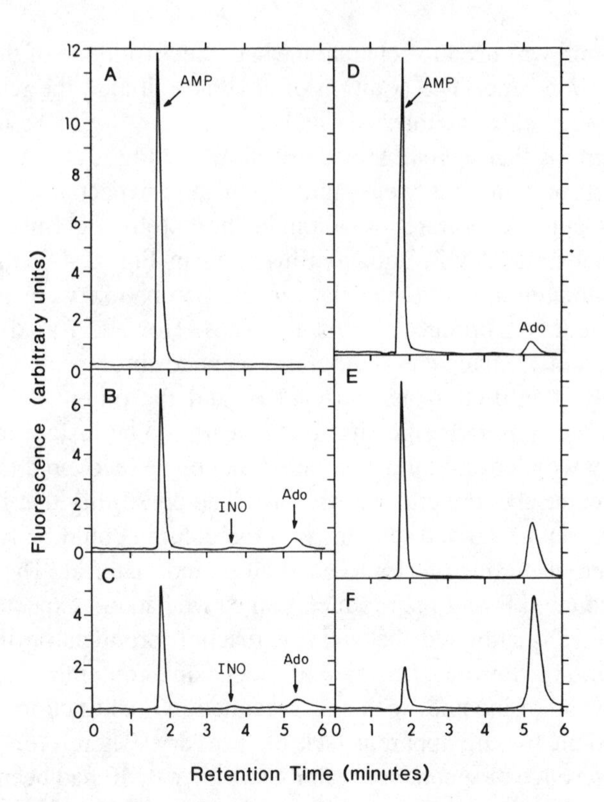

Figure 3.4 The hplc analysis of a reaction mixture containing AMP and alkaline phosphatase. Separations were carried out on a reverse-phase column with a mobile phase of potassium phosphate (pH 5.5) and 10% methanol. The column was eluted isocratically, and the detection was at 254 nm. Tracings obtained of reaction mixture with no EHNA: (A) immediately after the addition of enzyme, (B) 10 min; (C) 15 min. Arrows indicate elution position of inosine and adenosine. Panels (D)–(F) represent tracings obtained of reaction mixture containing 5 μM EHNA. Samples taken and analyzed (D) after 2 min, (E) after 10 min, and (F) after 40 min. (From Rossomando *et al.* 1981)

ample, if an analog of the substrate is used, then an analog of the product will be formed. If the latter is not a suitable substrate for the secondary enzyme, then no secondary reactions will occur. Alternatively, one can try to adjust the reaction condition in such a way that the enzymes catalyzing the secondary reactions will not be active. For example, if the primary reaction does not require metals but the secondary reaction does, adding a chelator will inhibit the latter.

Table 3.2 Secondary Reactions

What are they?

Secondary reactions are the result of enzymatic activities present in the sample that lead to destruction of the substrate and/or the formation of additional products, which can alter the appearance of the chromatogram.

Avoiding them

1. Purify the enzyme to homogeneity.
2. Use analogs whose products are not substrates for secondary reactions.
3. Adjust reaction conditions to minimize activity of secondary reactions.

3.1.6 Components of the Reaction Mixtures Can Cause Problems: Effects of Metals on Separation

Many enzymes require metals for activity, and, unfortunately for hplc use, the presence of the metal can occasionally have significant effects on the separation. For an explanation of this problem, return to the reaction illustrated in Fig. 3.1, the degradation of ATP to form AMP and PP_i. The first hplc method developed to assay this activity was carried out on a reverse-phase system with a mobile phase chosen for the exclusive separation of ATP from AMP (see Fig. 3.5). Since ADP was not involved, no thought was given to its separation. Later, having decided to study the metal requirements of this reaction, a series of experiments were performed for this purpose. A reaction mixture was prepared that contained a metal at concentrations in excess of that of ATP. The reaction was started by the addition of the enzyme, and samples were taken and analyzed by the hplc method. Surprisingly, the chromatograms for the experiments that included metals were different from the ones obtained previously. In the previous experiments, only two peaks were present, those representing ATP and AMP. However, in the experiment where the metal calcium was included, the chromatograms showed that an additional peak was eluted just after the ATP emerged (Fig. 3.5B). Further studies showed that this peak had a retention time identical to that of ADP, and therefore we assumed that this second peak was ADP. From these findings, we speculated that the metal had stimulated the activity of an enzyme that catalyzed the formation of ADP.

Additional studies were performed. We analyzed samples of the reac-

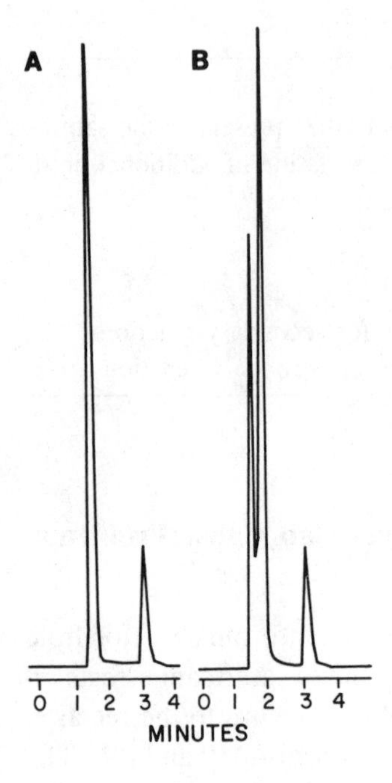

Figure 3.5 Hplc analysis of enzymatic assay with ATP in free and metal-bound forms. Separations were carried out on a reverse-phase C-18 column with a potassium phosphate mobile phase containing 10% methanol. The flow rate was 2 mL/min. Assay mixture of 100 μL contained 8 m*M* Tris-HCl (pH 7.5), 2 m*M* ATP, and 2 m*M* $MgCl_2$ and enzyme preparation containing ATP pyrophosphohydrolase (10 μg protein). Chromatograms of 20 μL samples illustrate incubation of (A) zero time and (B) 2 h. (From Jahngen and Rossomando, 1983).

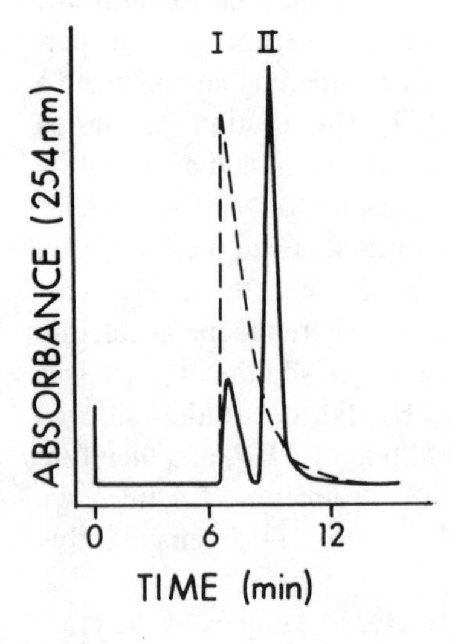

Figure 3.6 Hplc separation of ATP/Ca^{2+} mixture. Chromatographic conditions: C-18 (μBondapak) microcolumn (4.6 mm × 25 cm); mobile phase of 10 m*M* KH_2PO_4 (pH 5.5) with 4% methanol; flow rate of 0.5 mL/min.; 20 μL sample volume; detection at 254 nm. Dashed line, 40 nmol ATP; solid line, 40 nmol ATP plus 160 nmol $CaCl_2$. (From Jahngen and Rossomando, 1983).

tion mixture before the enzyme was added and before any enzymatic reaction could have taken place. Interestingly, these chromatograms, Fig. 3.6, also showed two peaks, with peak II identical in retention time to the presumptive ADP peak. In the absence of any metal, a single ATP peak (peak I) was observed, suggesting that the metal had altered the chromatographic properties of ATP. Additional studies confirmed this possibility. For example, when the effect of metals on the chromatographic behavior of ATP was studied in more detail, the profiles illustrated in Fig. 3.7A–F were obtained. Figure 3.7A shows the profile at the lowest metal concentration used in the experiment. Two peaks of ATP, labeled I and II, are clearly resolved. Increasing the metal concentration as shown in Fig. 3.7B–F results in an increase in the area of peak II and a corresponding decrease in peak I. This result suggests that the second peak was formed by the metal binding to the ATP.

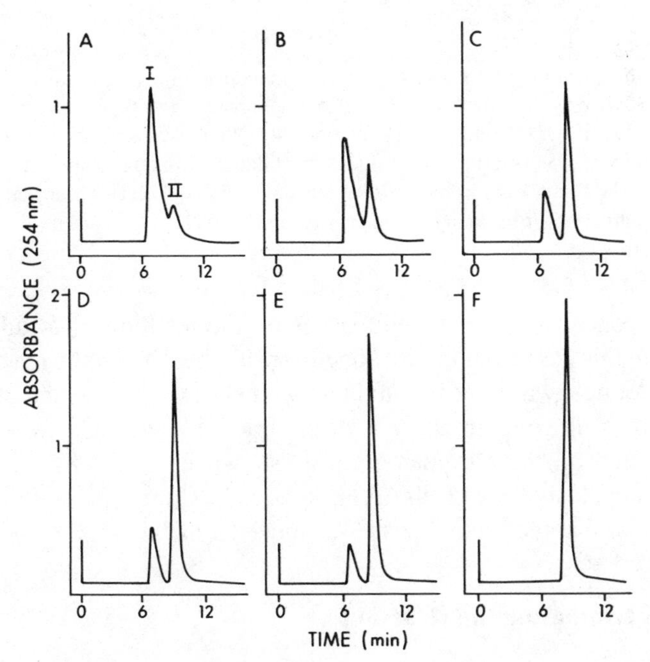

Figure 3.7 Effect of increasing amounts of Ca^{2+} in ATP/Ca^{2+} mixture. Analysis by hplc as described in Fig. 3.6. Chromatograms are of 20 μL samples containing 40 nmol ATP and $CaCl_2$ at (A) 60 nmol, (B) 100 nmol, (C) 120 nmol, (D) 160 nmol, (E) 200 nmol, and (F) 400 nmol. (From Jahngen and Rossomando, 1983).

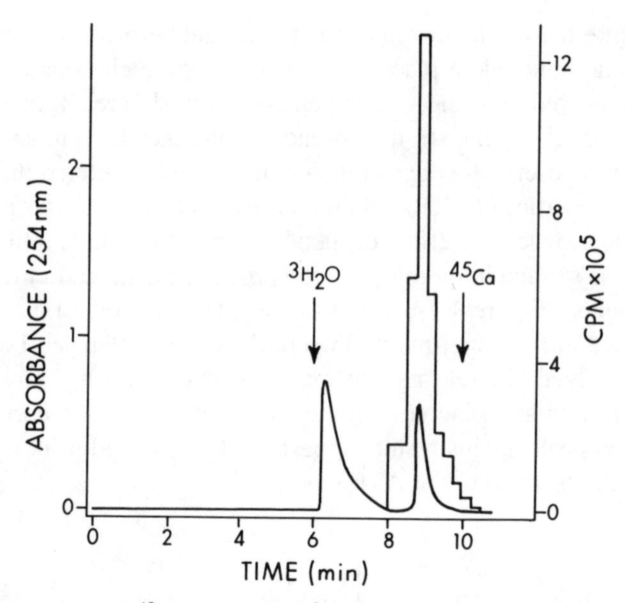

Figure 3.8 Coelution of ^{45}Ca with ATP/Ca^{2+} mixture. Injection volume and analytical conditions were as described in Fig. 3.6. Approximately 3 million cpm ^{45}Ca was added to 40 nmol ATP and 80 nmol $CaCl_2$. Fractions of 0.25 nL were collected, and the amount of radioactivity was determined by liquid scintillation counting. Radioactivity for ^{45}Ca with ATP/Ca^{2+} is shown in bars. Retention times of ^3H_2O and ^{45}Ca injected alone are shown by arrows. (From Jahngen and Rossomando, 1983).

This conclusion was tested further by the addition of a radiolabeled metal, in this case calcium-45, together with the ATP in the reaction mixture. A sample was injected, and the two peaks were collected in separate fractions. Following an analysis of the fractions, the ^{45}Ca elution profile was plotted on the chromatogram as shown in Fig. 3.8. The calcium coelutes exclusively with the ATP in peak II (Fig. 3.8), confirming the conclusion that the second peak was indeed a metal-ATP.

3.1.7 Terminating the Reaction

In designing an assay for an enzyme, it is often necessary to introduce a termination step into the protocol (see Chapter One). This is often the case when protein is present in the incubation mixture at a concentration that would clog the column. There are a variety of ways to accomplish this termination process. The ideal procedure would be one that did not add reagents that might otherwise clog the column or alter its performance. For example, consider the changes that occur in the incubation mixture when the reaction is terminated by acid. The addition of trichloroacetic acid (TCA) will reduce the pH of the incubation solution to a value unsafe for the stationary-phase packing. In addition, the TCA has an absorption profile in the UV region, and therefore its presence will interfere with the use of UV spectrophotometers as detectors. Both problems can be solved by removing the TCA prior to analysis. The freon-alamine extraction method introduced by Khym, Bynum, and Volkin (1977) has been found to be useful for this purpose. While the removal of TCA is not a difficult step, it is clear that its inclusion in the assay complicates the original procedure.

A method that offers an alternative to the use of acids and bases is the addition of chelators such as EDTA. This technique is suitable only for reactions in which the enzymatic activities have an absolute requirement for a metal whose removal will terminate the activities.

Another alternative method we have found useful for terminating reactions is to heat the incubation mixture to a temperature that results in inactivation of the enzyme. Usually temperatures in excess of 100°C are required. One of the techniques often used is to immerse the reaction tube in a bath of boiling water. In my laboratory, this method is not employed because the incubation mixture cannot be brought to 100°C quickly enough to effect instantaneous termination.

We have tried using commercially available heating blocks. Such heaters were also found unsuitable, for the rather trivial reason that test tubes being used for the incubation did not fit the holes in the block.

Finally, a simple device—a sand bath—was found effective in terminating reactions instantly. We filled a stainless steel rectangular pan (about 8 in. × 10 in.) with about 2 in. of sand and placed it on a hot plate. This setup is illustrated in Fig. 3.9A. The temperature of the sand bath is easily brought to 155°C, and this temperature can be maintained throughout the working day without fear of evaporation. There is never a problem of fitting the tubes—one simply thrusts any size tube directly into the sand. The insertion of an incubation tube containing as much as 500 μL of incubation mixture resulted in the temperature inside the solution reaching 100°C "instantly," thus terminating the reaction.

Termination of most enzymatic reactions with heat results in precipita-

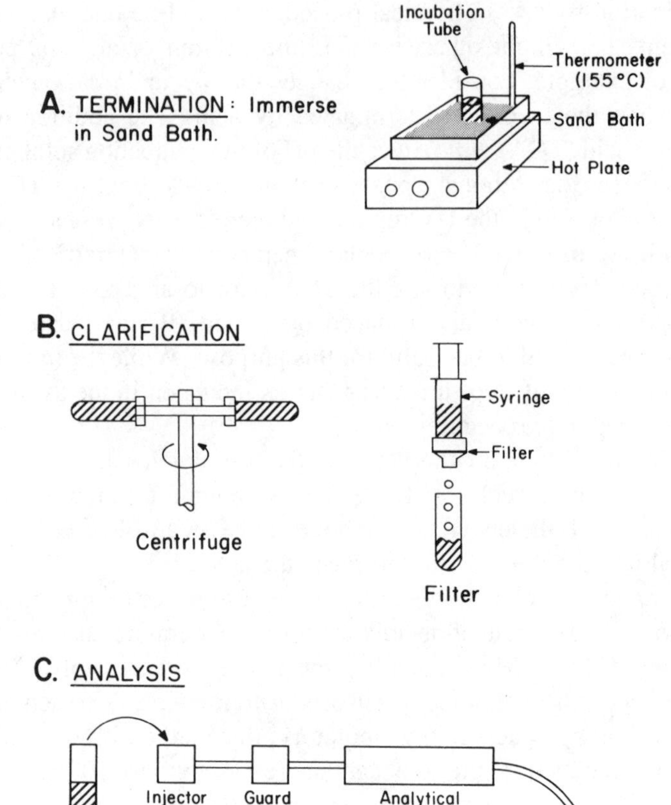

Figure 3.9 The preparation of a sample from a reaction mixture that contains an excess amount of protein prior to injection and analysis by hplc. (A) Termination carried out by immersion of reaction sample in a sand bath maintained at 155°C; (B) removal of the precipitated protein by either centrifugation or filtration; and (C) injection and analysis of the clarified solution.

tion of any proteins present in the reaction mixture. Because this precipitation is irreversible, and because when crude extracts are being assayed the amount of protein may be considerable, it is often necessary to remove the precipitate prior to sampling.

The precipitate is removed either by filtration or by centrifugation as illustrated in Fig. 3.9B. Because of the small volumes usually used in the reaction mixture, volumes between 100 and 500 μL, centrifugation is difficult. Recently, filters with dead volumes of about 50 μL have been developed that make the removal of precipitate less tedious. Following the removal of the precipitate, a sample may be removed from the filtrate and injected into the hplc for analysis as illustrated in Fig. 3.9C.

Internal standards, compounds added at any stage of the analtyical procedure, can be useful in calibrating and/or calculating the effect of that procedure on the recovery of the substrate or product of the reaction. The compounds chosen as internal standards should elute close to the substrate or product and have similar detection characteristics.

3.1.8 Setting Up the Reaction Conditions

Some additional modifications of the reaction conditions used with other assay methods may be required in order to proceed with the hplc assay method. One change is related to the total volume of the reaction mixture. Because the hplc assay is basically a discontinuous technique, obtaining kinetic data requires multiple samples, each one representing a single time point. Traditionally, reactions requiring multiple sampling have been arranged in one of two ways. In one arrangement, illustrated in Fig. 3.10A, separate reaction mixtures are set up, each one representing a single time point. In this case, the total volume required for a single reaction mixture would be the volume required for a single injection. The number of incubation tubes would be determined by the number of time points required by the experiment. In the second arrangement, shown in Fig. 3.10B, a single incubation mixture is prepared, and samples are removed from it at suitable intervals for analysis. In this arrangement, the volume required for the reaction mixture would be determined as the product of the volume needed for each injection multiplied by the total number of injections.

Since with both arrangements the volume of a single injection is the important variable, it would appear that once this value is determined the overall reaction volume can be established. However, another variable, the type of injector to be used, must also be considered.

Injectors are of two basic designs. Those of the type illustrated in Fig. 3.11A require that a sample of known volume be removed from the reaction and injected. Thus, with this type of injector, the volume to be injected is determined by the volume drawn into the syringe.

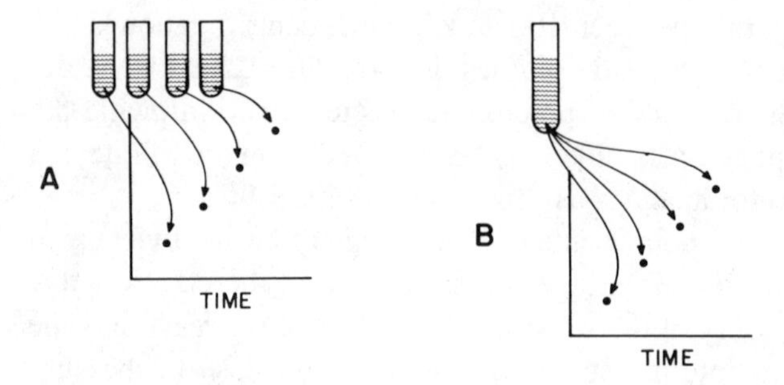

Figure 3.10 Representation of two procedures used to obtain multiple samples for analysis. (A) In this arrangement, several identical reaction mixtures are prepared, and the enzyme will be added to each to start the reaction. During the incubation each tube will be sampled only once. (B) In this arrangement, only one reaction mixture is prepared and the enzyme is added to start the reaction. The incubation mixture is sampled repeatedly during the course of the reaction. Note that the volume of the reaction mixture in arrangement (B) is usually greater than in arrangement (A).

In the second type, the injector unit contains a device called a "loop" (Fig. 3.11B), which is a coiled tube of precise volume. The loop is loaded with sample, and once filled, its contents are injected onto the column. It is always necessary to overfill such a loop to ensure complete loading. To overfill, the loop is loaded with sample until it overflows; the excess sample spills out one of the injection valve vents. Although injectors differ in the amount of overfilling required, our experience would recommend overfilling by as much as 30%, since it is critical to load the injector loop completely. Partially filled loops will result in abnormal chromatographic profiles. There is waste with injectors of the type shown in Fig. 3.11B, and if material is in short supply or expensive, the type of injector shown in Fig. 3.11A might be preferred.

3.1.9 Detector Sensitivity

One more potential problem should be discussed. It concerns the question of selecting the range of substrate concentrations to be used throughout the study. Considering the sensitivity of most hplc detectors and the apparent K_m values of most enzyme activities, the selection of the upper

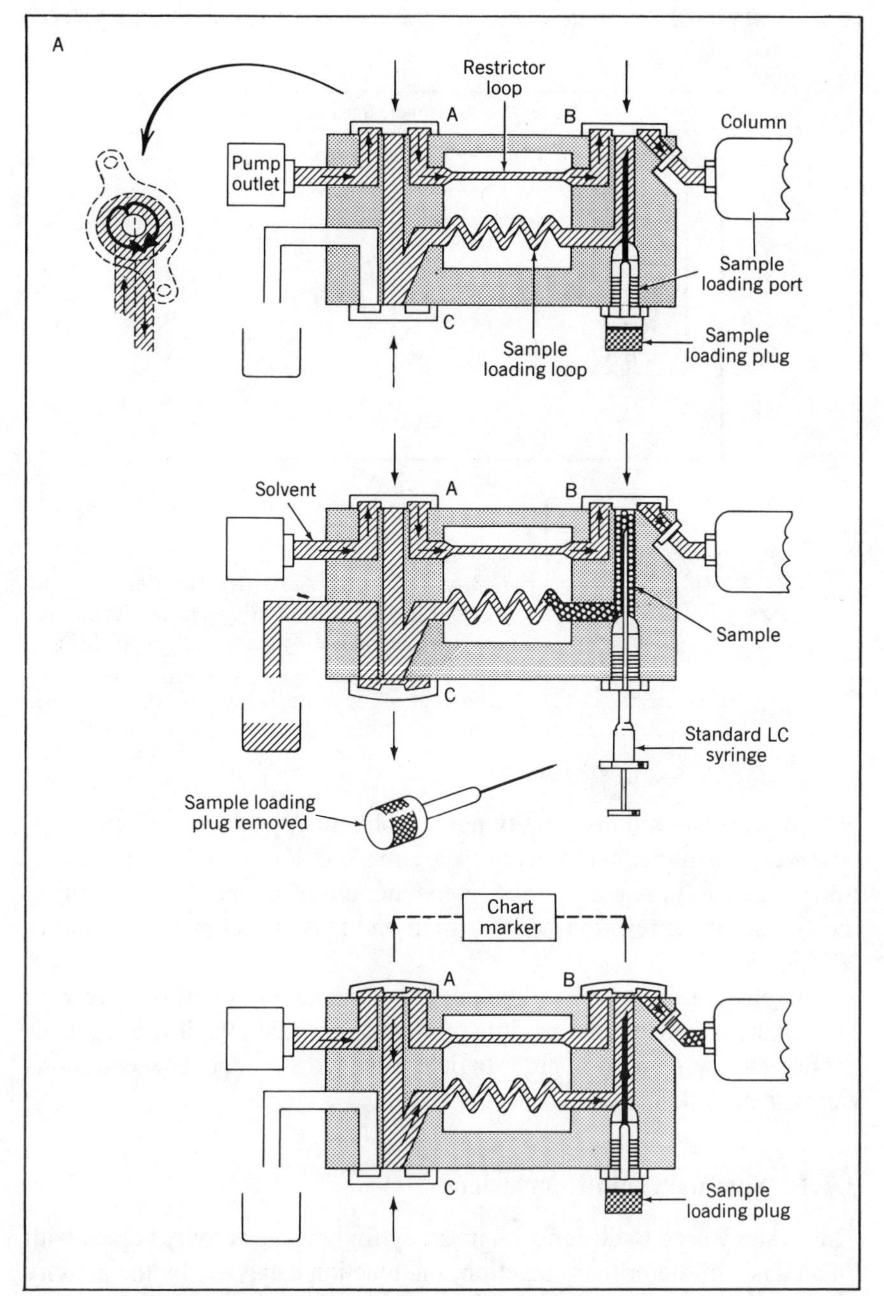

Figure 3.11 High-pressure injection valves. (A) The universal injector (Waters, Inc.). A known volume of sample is injected, displacing a corresponding volume from the "sample loading loop" shown in top diagram. After sample loading plug is replaced, loading loop is placed "in line," and eluent from pump moves sample onto column.

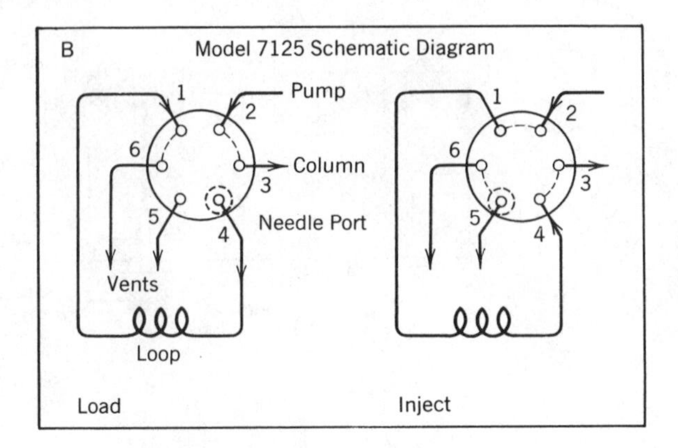

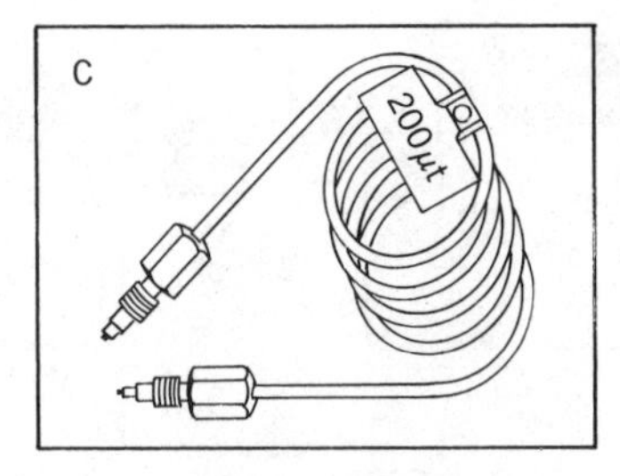

Figure 3.11 (Continued) (B) The Rheodyne injector (Rheodyne, Inc.) in which the volume injected is determined by the sample loop. Sample loops can be obtained commercially in a number of fixed volumes or made to any size. (C) A 200-μL loop.

limit of concentration is usually not a problem. A problem will develop, however, when rate determinations are made at low substrate concentrations, since at these concentrations the amount of product formed during the course of the reaction will be small and may be below the monitor's level of detection.

Therefore, prior to executing any experimental protocols dealing with low substrate concentrations, it is prudent to ascertain the lower limits of the detector being used in order to determine what product concentrations can be detected.

3.1.10 Summary and Conclusions

The strategy used to design an hplc assay for enzyme activity begins with an analysis of the primary reaction, the reaction catalyzed by the activity of interest. (Figure 3.12 shows the steps in the assay design.) This step should be followed by an analysis of secondary reactions. These are reactions catalyzed by activities that might be present in the sample and

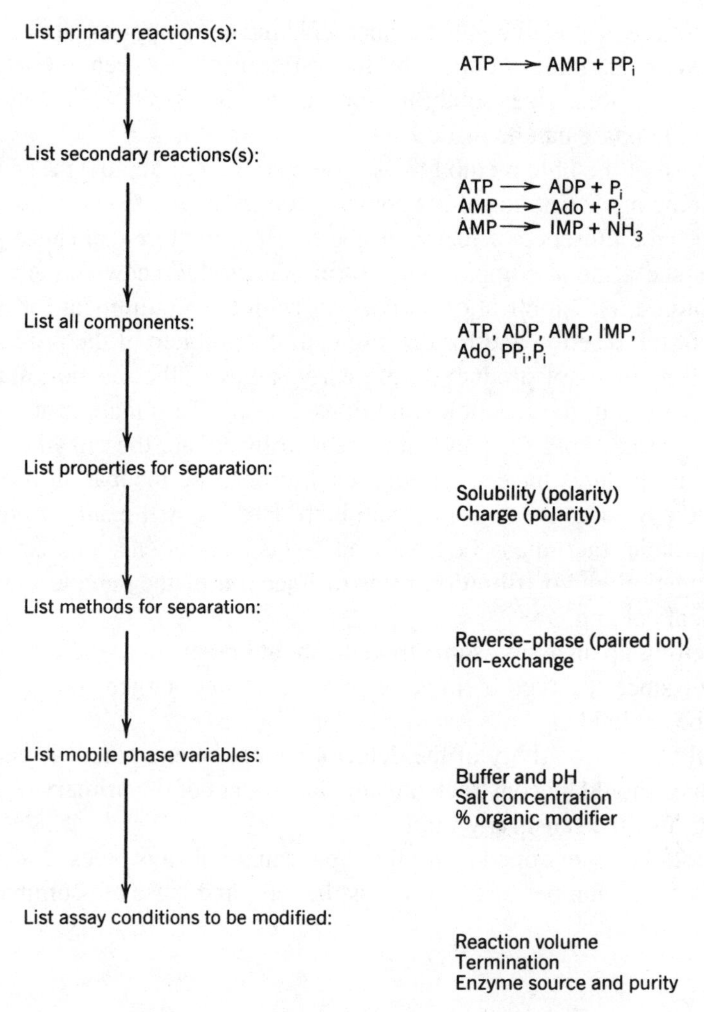
List primary reactions(s):
ATP ⟶ AMP + PP_i
List secondary reactions(s):
ATP ⟶ ADP + P_i
AMP ⟶ Ado + P_i
AMP ⟶ IMP + NH_3
List all components:
ATP, ADP, AMP, IMP,
Ado, PP_i, P_i
List properties for separation:
Solubility (polarity)
Charge (polarity)
List methods for separation:
Reverse-phase (paired ion)
Ion-exchange
List mobile phase variables:
Buffer and pH
Salt concentration
% organic modifier
List assay conditions to be modified:
Reaction volume
Termination
Enzyme source and purity

Figure 3.12 Steps in the design of an hplc assay.

are not of primary interest but the presence of which might result in loss of substrate or products and thereby obscure the results of the primary reaction.

The choice of the stationary phase can best be made after the analysis of both the primary and secondary reactions is complete and a list of those compounds to be separated has been obtained. The selection of the stationary phase is guided by the number and kind of compounds in the reactions. With an understanding of the differences between these compounds, be it their size, solubility, or charge, the selection of stationary and mobile phase can be made.

The use of the hplc method to assay the activity of an enzyme may require some modifications in the composition of the reaction mixture. For example, the presence of metals in the reaction mixture can cause problems, since a metal complex may form and produce new peaks on the chromatogram. Complications associated with the requirement for termination of the reaction and for dealing with the problem of the production of small amounts of product during early stages of the reaction may require changes in the reaction conditions as well. Termination of a reaction is best accomplished by using heat to inactivate the enzymes. Numerous procedures have been employed to heat the incubation mixture; one of the most convenient is a sand bath. Heating will result in precipitated protein that must be removed. This can be accomplished by centrifugation and/or filtration prior to injection of the sample onto the analytical column.

In setting up an assay, some thought should be given to the incubation volume, since the type of injector may necessitate larger volumes than might be available.

Finally, the sensitivity of the detector should be examined. A calibration curve should be constructed using the product of the primary reaction to determine the sensitivity limit.

It should be mentioned again that hplc cannot always be used to assay an activity. A number of criteria must be met, and these are summarized in Table 3.3.

Table 3.3 Conditions that Allow the Use of HPLC for the Assay of an Enzymatic Activity

1. A system (stationary and mobile phase) can be found for the separation of substrate(s) from product(s).
2. A detection system for substrate(s) or product(s) is available.
3. Suitable standards exist for calibration and identification of reaction components.
4. Sufficient product is formed to follow the course of the reaction.

3.2 THE USE OF HPLC TO ESTABLISH OPTIMAL CONDITIONS FOR THE ENZYMATIC REACTION

3.2.1 Initial Decisions: Composition of the Reaction Mixture

The hplc assay method is particularly useful when it is necessary to obtain the initial rate data necessary to the study of an enzymatic activity. Optimal assay conditions for the hplc must be established first. Usually, the optimization process involves the determination of several variables, such as the optimal substrate concentration, pH, temperature, and enzyme concentration. It will be assumed that the reader is familiar with the problems associated with assay conditions such as pH, buffer, and temperature. In this chapter only those factors that might present problems for the hplc assay method will be discussed. For additional information see the works cited in the General References.

When studying an enzymatic activity for which an apparent K_m value can be obtained from the literature, the determination of the optimum substrate concentration becomes somewhat easier. Thus, a concentration of 2–3 times the K_m value (assuming the absence of "substrate inhibition") is adequate for use in early experiments. The literature can provide information on other parameters, such as the pH range, the requirement for activators, and the optimal temperature for the incubation. Armed with this information, and with the incubation conditions determined, the reaction mixture can be prepared.

What remain to be determined prior to the initiation of the reaction by the addition of the enzyme are the amount of enzyme to be added to the reaction mixture, the time course of the reaction, the time between samplings of the reaction mixture, and the volume of these samples. All four questions can be answered by trial and error using the following scheme. First, an arbitrary enzyme concentration is selected. While any concentration can be used, it should be remembered that when working with a crude extract, excess protein can clog the column. Therefore, choose the lowest concentration possible.

Having chosen an enzyme concentration, add this amount to the reaction mixture to start the reaction and sampling of the reaction mixture can begin at any time, by the withdrawal of a predetermined quantity for analysis.

The chromatogram obtained from this single sample is examined for a new peak—the product. Two outcomes are possible. Either product is present or not. If a product peak is detected, and its area is very small compared to the area of the substrate peak, then a second sample can be withdrawn from the incubation mixture and injected for analysis.

Again the areas of the substrate and product peaks should be compared. If the area of the product peak is more than 50% of that of the substrate, the reaction has progressed too far, and it is necessary to start again by preparing a new reaction mixture. In order to obtain more time points, the reaction rate should be slowed by using less enzyme.

Alternatively, in the absence of the formation of any product, incubation should be continued and samples should be withdrawn every hour and analyzed. The incubation can be continued for several hours until product is detected. In the absence of any detectable product, a new reaction mixture can be prepared which contains more enzyme than the first. If this does not result in the formation of detectable product, the possibility should be considered that the fraction being assayed contains no activity.

3.2.2 Obtaining Initial Rate Data

The significance of obtaining initial rate data for the study of enzymes has been discussed elsewhere, and the reader is referred to the General References for additional information. Usually, such concerns are relevant to the in-depth study of the mechanism of an enzyme reaction and beyond the scope of the present discussion. Of concern in this text are the problems associated with obtaining initial rate data with the hplc assay method.

As a result of the experiments described above, values will have been obtained for two parameters: the amount of the enzyme required to form sufficient detectable product and the incubation time required to form this amount of product. Additional experiments will now be required to refine the values of both parameters. Keep in mind that at least three time points will be required to generate the straight line needed to obtain the initial rate. To obtain these points, proceed as follows: First, note that if the rate of product formation is too rapid—that is, the reaction rate becomes nonlinear before three or four samples can be analyzed—then the rate should be slowed by decreasing the amount of enzyme. Alternatively, if the rate of the reaction is too slow, so that it takes all day to form product, the enzyme concentration should be increased so that three or four samples may be analyzed in about 2 h. In the absence of terminating a reaction, the overriding parameter that governs the sampling interval is the time needed to complete the hplc analysis of each sample. It will not be possible to inject a second sample until the first has been completely eluted and the column prepared for the second sample. If this elution and preparation takes 15 min, then it is this value that establishes the minimum sampling time. Thus, the time required for the hplc analysis of a single sample will determine the concentration of the enzyme used in the reaction. Once a suitable concentration of enzyme has been established so that three or four samples can be analyzed, then the quantitative data can be obtained. A reaction is started, the reaction mixture is sampled at intervals, chromatograms are obtained, and the amount of product formed is determined directly from the chromatogram using either peak height or electronic integration of the peaks. These values should be plotted as a function of reaction time.

Next, a second and third series of reaction mixtures should be prepared, with enzyme added at concentrations of one-half and twice the value used in the first. These reactions are started and sampled, chromatograms are obtained, and the data plotted as a function of reaction time. It should be noted that at this early stage in the optimization of the assay it is advisable to continue sampling one of the incubations until the rate of product formation becomes nonlinear or the amount of substrate present is exhausted. This prolonged incubation provides information about the extent of the primary reaction and also allows any secondary reactions to take place and form enough products to be detectable.

3.2.3 Quantitative Analysis of the Reaction

The chromatograms used to obtain the initial rate data described above may be examined to provide some information on the fate of the substrate during the course of the incubation. For example, assuming no other reactions involving the substrate have taken place, the amount of product that was formed should be equal to the amount of substrate that was lost. It is important to determine this point first.

A careful visual inspection of the chromatogram should be sufficient to note the presence (or absence) of any peaks other than the substrate and the product. If only the latter is present, this would suggest the absence of secondary reactions involving either substrate or product. Second, a vi-

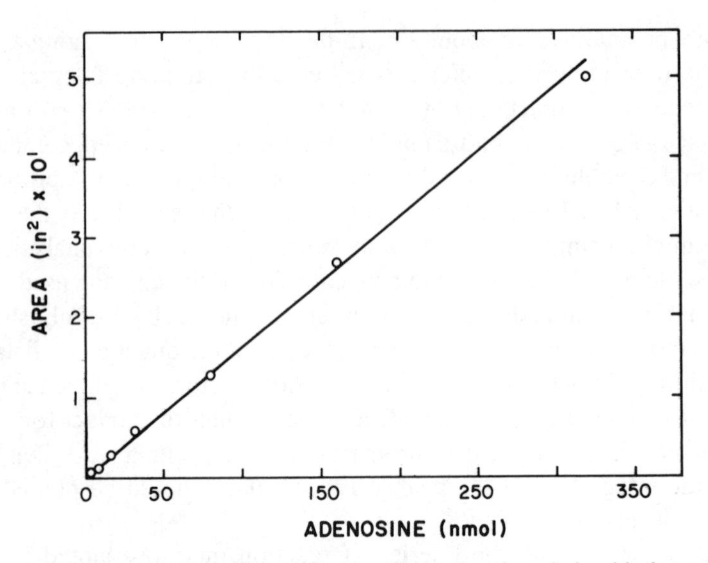

Figure 3.13 Representative calibration curve illustrating the relationship between peak area and the concentration of adenosine.

sual estimation of the area of both the substrate and product peaks should indicate whether the two are equal assuming an equivalence or their extinction coefficients. Estimations will be more difficult if the widths of the substrate peak and the product peak differ.

It is also useful at this stage to become more quantitative by converting the amount of product formation from the "machine units," arbitrary integration units or percent obtained, directly to units of amount. Such a conversion requires access to a calibration curve that relates the machine units to more specific units of amount. An example of such a calibration curve for adenosine is shown in Fig. 3.13. Having made the conversion, the initial rate of product formation determined above can now be plotted as a function of enzyme concentration as part of the optimization process.

As a result of these procedures, a graphical representation of the rate of product formation will be obtained. Such data can be analyzed visually or be subjected to statistical analysis (see General References for details). In addition, data collected at several substrate concentrations can be used to obtain kinetic constants. Again, these data, collected by the hplc method, can be manipulated by standardized methods (see General References).

Obtaining initial rate data is, of course, a first step in the kinetic analysis of an enzyme-catalyzed reaction, and the reader is referred to the General References at the end of the chapter for several reviews and monographs describing the methods for this analysis.

For the two substrate–two product reactions, the Bi Bi reactions in Cleland's nomenclature, the hplc assay method is particularly useful. Its application, however, will require that methods be developed for the separation of the two substrates and the two products. Initial rate data generated by the application of the hplc assay method can be used to produce double-reciprocal plots [1/(initial velocity) versus 1/(substrate concentration)], which can in turn be used to differentiate between sequential and ping-pong reaction mechanisms. Thus, as described by Alberty (1953), for a sequential mechanism the several lines on such a plot will converge or intersect at a common point, whereas for the ping-pong mechanism the series of lines are nonconvergent or parallel.

To date, the application of the hplc assay method to studies on reaction mechanisms has been limited, and the reader is referred to the chapter by Sloan (1984). Sloan and his colleagues studied the formation of IMP or GMP (and pyrophosphate) from the substrates phosphoribosylpyrophosphate (PRPP) and either hypoxanthine or guanine. These reactions, catalyzed by hypoxanthine-guanine phosphoribosyltransferase (GHPRTase), were studied by hplc after a method was developed to separate all the reactants and products simultaneously.

The strategy used was as follows: First the initial rates of IMP and GMP formation were studied separately, and from an inspection of the double-reciprical plots obtained from these data it was determined that product formation proceeded by a sequential kinetic mechanism. Next the formation of IMP and GMP was studied for several concentrations of both bases, and, finally, the rate of formation of the common product, pyrophosphate, was examined over a series of fixed hypoxanthine/guanine ratios.

On the basis of a graphical analysis of these data, it was concluded that the reaction proceeded by an ordered Bi Bi mechanism. According to this mechanism, first the substrate PRPP would bind to the enzyme and then either base would bind to the same site on the complex.

While such a study might have been carried out with conventional methods, the use of hplc facilitated this study considerably by allowing all reactants and products to be measured in one analysis. Clearly, the hplc assay method should be considered when the kinetics of a multisubstrate enzyme reaction are to be studied.

3.2.4 Initial Rate Determination at Low Substrate Concentrations

As in the case of any assay procedure, the determination of the rate of product formation becomes difficult at the lower limits of substrate concentration. However, there are changes that can be made both in the assay system and in the chromatographic equipment that can solve this problem.

The first change is, of course, to increase the sensitivity of the detector. Most hplc detectors contain range switches that make this a simple matter. When range switching is carried out, it is useful to determine if the calibration curves constructed at one range setting are still valid at another.

Next, the amount of product being detected may be increased by increasing the volume of the reaction mixture that is injected for analysis. With fixed loop injectors this requires changing the loop, a rather simple procedure. When the loop is changed it is necessary to consider the additional volume when calculating the total volume of the incubation mixture.

There is an upper limit to the volume of an injection, usually around 200 μL. The injection of larger volumes results in spreading of the peak, a phenomenon that decreases resolution. However, it is possible to make loops of any volume and thus control this upper-volume limit. If separation is adequate and resolution is not a problem, the loop volume may be increased.

There have been situations in which obtaining enough reaction product requires the concentration of an entire reaction mixture. Following concentration, the residue is resuspended in a small volume of buffer and analyzed.

Finally, it is always possible to increase sensitivity by using analogs, such as radiochemicals, as substrates and determine the amount of radioactive product that has formed. For experiments of this type, the eluent may be carried through a radiochemical detector; in the absence of such an instrument, fractions may be collected and their radioactive content determined.

3.2.5 The "Sensitivity-Shift" Procedure

In most enzymatic reactions it is not uncommon for a sample to contain a concentration of substrate 100- to 1000-fold greater than that of the reaction product. This situation often arises with enzymes that require high concentrations of substrate (low K_m) for reactivity and have low rates of product formation. It also occurs soon after the start of the reaction when only small amounts of product have been formed.

The analysis of a sample from an incubation mixture that contains significant differences in substrate and product concentrations presents some problems, since on one hand the detector must be set so that the substrate peak is on scale, and on the other it must be set to detect small amounts of product. Usually different sensitivity settings are required to put both components on scale. For these cases we have adopted a procedure that might be called the "sensitivity shift." In this procedure, the injection is made with the sensitivity set at a value that allows for the detection of one of the compounds. If this is the product, the detector is set at its maximum sensitivity. Following the emergence of the product, the sensitivity of the detector is immediately changed (either manually or electronically by computer) to a value that allows the substrate peak to appear completely on scale.

Assuming that the settings on the detector are proportional, there should be no difficulty in relating the areas of both peaks to the calibration curve. These problems can be avoided with electronic integrators which measure total absorbance and are not affected by sensitivity settings.

3.2.6 Substrate Analogs: Their Use in Limiting Secondary Reactions

The presence of enzymes that catalyze unwanted secondary reactions is a problem that hplc users must avoid. In cases where secondary reactions make it difficult to quantitate the primary reaction, one solution is to use an analog as the substrate for the latter. The analog should be chosen such that while it is a substrate for the primary reaction the product formed is not a substrate for the secondary reaction.

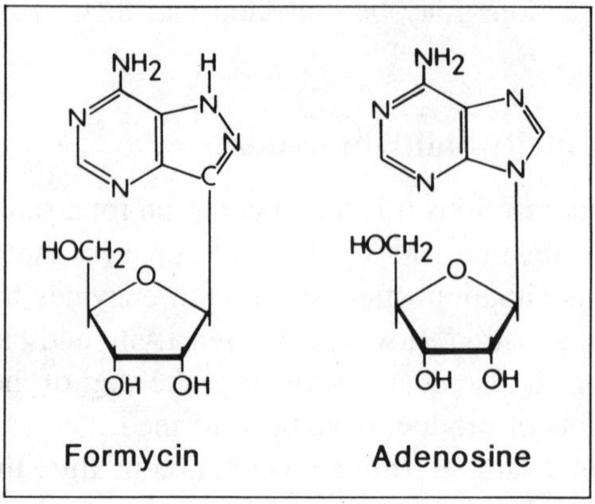

Figure 3.14 Comparison of structures of adenosine and its fluorescent analog formycin A.

Analogs can be used in another way. Consider the case of developing an assay procedure for adenosine kinase, the enzyme that catalyzes the primary reaction Ado + ATP → AMP + ADP. Problems will arise during the assay of this activity in crude extracts, since other enzymes may be present that can form AMP directly from ATP.

Radiochemical analogs such as radiolabeled adenosine are ideal for solving this problem, because if the formation of radiolabeled AMP is monitored, it is possible to distinguish the AMP formed from adenosine from that formed from ATP, which, of course, would not be labeled.

Alternatively, this same reaction can be assayed if adenosine is replaced by formycin A (FoA) (Fig. 3.14) a fluorescent analog. With this substrate, one product of the adenosine kinase reaction would be FoMP, the fluorescent analog of AMP, while AMP formed directly from ATP would not be fluorescent. Therefore, monitoring of both the fluorescence and the ultraviolet absorbance, using equipment arranged as shown in Fig. 3.15, would make it possible to follow the progress of both the kinase reaction and any secondary reactions.

3.2.7 Summary and Conclusions (Fig. 3.16)

Prior to initiation of the reaction, the concentration of the substrate and other components of the reaction mixture must be established. Decisions

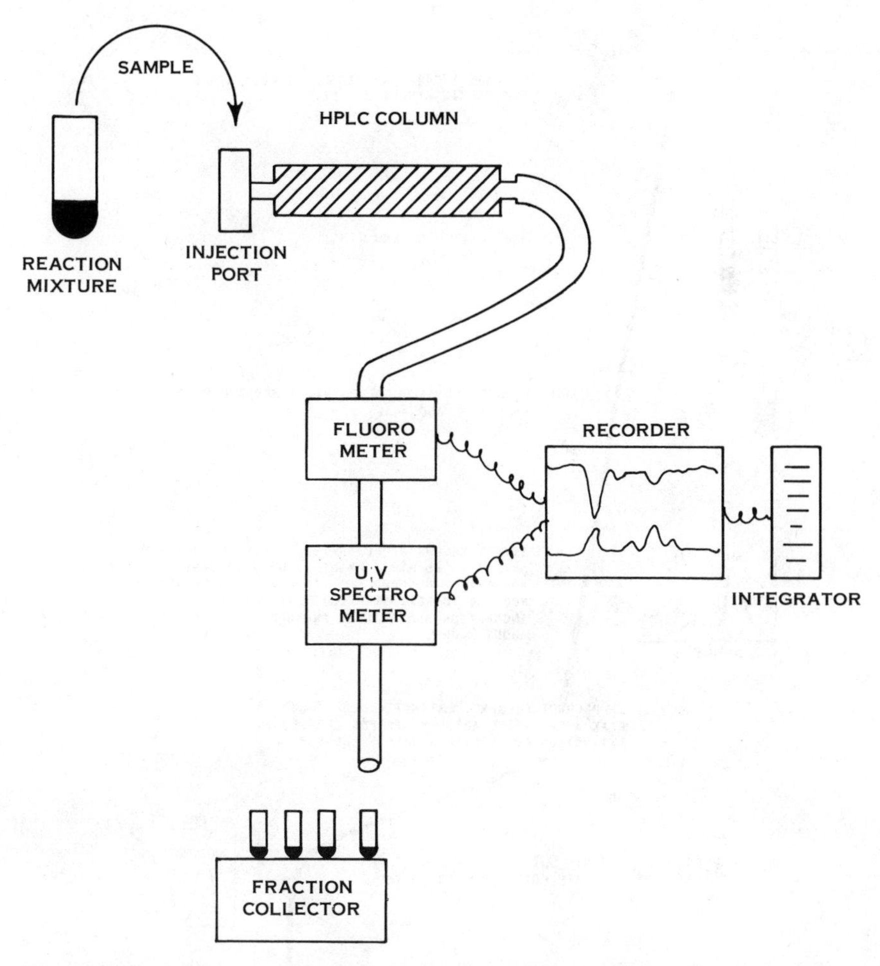

Figure 3.15 Arrangement of equipment showing relative position of multiple detectors and fraction collector.

regarding enzyme concentration and sampling intervals must be made empirically with some information about the amount of expected product formation. These data will provide limits for the enzyme concentration and incubation period. Also, if a reaction is not to be terminated, the time required to elute a single sample must be established. This value will also provide an estimate of the minimum interval between samples.

To obtain initial rate data, it is necessary to have a minimum of three

SETTING-UP AN ASSAY

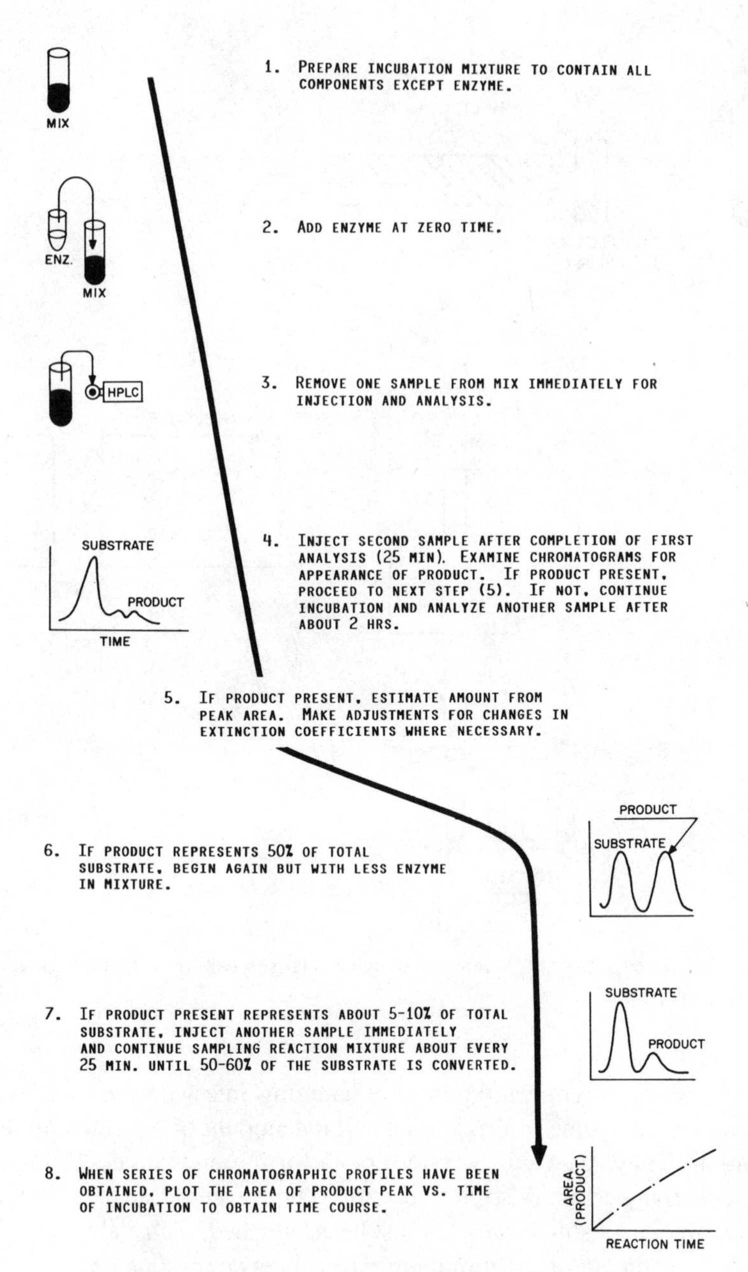

Figure 3.16 Strategy for optimizing an enzyme assay.

samples of a reaction carried out under optimized conditions. The formation of product at several enzyme and substrate concentrations will yield data for the calculation of kinetic constants.

The chromatogram obtained from each analysis provides information about product formation and loss of substrate. In the absence of secondary reactions the two values should be additive.

The use of the hplc method provides some unique solutions to the problem of determination of product formation at low substrate concentrations. For example, the sensitivity of the detector can be enhanced even during the course of an analysis. Next, the volume of the incubation mixture assayed may be increased to provide more product, and finally, it is possible to use analogs such as radiolabeled or fluorescent compounds for which there is greater detector sensitivity.

REFERENCES

Introduction to Kinetic Mechanisms

Alberty, R. A., *J. Amer. Chem. Soc.* **75**:1928 (1953).

Terminating Reactions

Khym, J. X., Bynum, J. W., and Volkin, E., *Anal. Biochem.* **77**:446 (1977).

Determination of Kinetic Mechanisms by the HPLC Assay Method

Sloan, D. L., Kinetic analysis of enzymatic reactions using high performance liquid chromatography, in J. C. Giddings, E. Grushka, J. Cazes, and P. R. Brown, eds., *Advances in Chromatography,* Vol 23, Dekker, New York, 1984.

GENERAL REFERENCES

General

Brown, P. R., *J. Chromatogr.* **52**:257 (1970).

Scoble, H. A., and Brown, P. R., Reversed-phase chromatography of nucleic acid fragments, in C. Horvath ed. *High Performance Liquid Chromatography, Advances and Perspectives,* Vol. 3, Academic, New York, 1983.

Enzymes and Reaction Conditions

Bergmeyer, H. N., Gawehn, K., and Moss, D. W., *Methods of Enzymatic Analysis,* Vol. I, VCH, Deerfield Beach, FL, 1974.

Dixon, M., Webb, E. C., Thorne, C. J. R., and Tipton, K. F., *Enzymes,* Academic, New York, 1979, Ch. 2.

Secondary Reactions

Palmer, T., *Understanding Enzymes,* 2nd ed., Wiley, New York, 1985.

Rossomando, E. F., Cordis, G. A., and Markham, G. D., *Arch. Biochem. Biophys.* **220**:71 (1983).

Rossomando, E. F., Jahngen, J. H., and Eccleston, J., *Anal. Biochem.* **116**:20 (1981).

Metals in Enzyme Reactions

Jahngen, J. H., and Rossomando, E. F., *Anal. Biochem.* **130**:406 (1983).

Introduction to Kinetic Mechanisms

Cleland, W. W., Steady state kinetics, in P. D. Boyer, ed., *The Enzymes,* 3rd ed., Vol. 2, Academic, New York, 1970.

Cleland, W. W., *Biochim. Biophys. Acta* **67**:104 (1963).

Fersht, A., *Enzyme Structure and Mechanism,* 2nd ed., Freeman, New York, 1985.

Fromm, H. J., Summary of kinetic reaction mechanisms, in D. L. Purich, ed., *Methods in Enzymology,* Vol. 63A, Academic, Orlando, FL, 1979.

Statistical Analysis of Initial Rate Data

Cleland, W. W., Statistical analysis of enzyme kinetic data, in D. L. Purich, ed., *Methods in Enzymology,* Vol. 63A, Academic, Orlando, FL, 1979.

Initial Rate Determinations

Fromm, H. J., *Initial Rate Enzyme Kinetics,* Springer-Verlag, Berlin, 1975.

Rudolph, F. B., and Fromm, H. J., Plotting methods for analyzing enzyme rate data, in D. L. Purich, ed., *Methods in Enzymology,* Vol. 63A, Academic, Orlando, FL, 1979.

Chapter Four

Strategy for the Preparation of Enzymatic Activities from Tissues, Body Fluids, and Single Cells

Overview

In developing a strategy for the preparation of an enzymatic activity it is useful to consider two factors, the first of which is the choice or selection of the biological samples to be used as the starting point for the purification. These samples will clearly differ in terms of their complexity, and it is this complexity that can be used to subdivide the samples into groups.

In the first group are those samples which are rather complex in that they not only consist of many different cell types but also have an extracellular compartment. Such samples include organs, tissues, biological fluids, and microbial cells, together with any other unicellular organisms grown in a culture medium or fermentation broth. For these samples, the initial step is the separation of the cellular from the noncellular compartment. Next, the different cell types within the cellular compartment must be separated, and this homogeneous population of each type of cell becomes the starting point for samples in the second group. With such cells, those activities at the cell surface can be directly assayed or the cells can be lysed, providing accessibility to the activities in intracellular organelles and on cytoplasmic fragments.

The samples in the third group are the subcellular fragments liberated by lysis. These include organelles such as mitochondria, as well as those operationally defined as a "membrane fraction" or a fraction containing "soluble components." The initial steps involving samples within this group include the separation of organelles from each other, the separation of insoluble from soluble fractions, the solubilization of membrane samples or the fractionation and separation of one molecular species from another. Since the strategy developed for purification depends on the choice of starting material, some of the problems associated with obtaining activities from samples in each of the groups are outlined in this chapter. In addition, solutions to these problems will be offered.

There is a second consideration involved in the development of a purification strategy. Derived in part from the first, this consideration relates to the question: To what extent should the activity be purified? The traditional end point of any purification scheme would be a homogeneous protein, because it was originally demonstrated that an enzymatic activity was associated with a single protein molecule. Thus, this question may appear to have only one answer. Several considerations can be used to justify this as the end point of the purification, the most important of which relates to the difficulties associated with the assay of the activity when the preparation is not homogeneous, for example, when the enzyme remains in a preparation that contains many proteins and many enzymes.

The development of the hplc method to assay enzyme activities has made it considerably easier to assay a single activity in the presence of others. Thus, attempts to obtain a pure protein during the purification procedure may not be necessary. Since the advent of hplc to assay enzyme activities, it is possible to stop the purification at a much earlier stage and still assay for a single enzymatic activity. In fact, for some studies, it is even advantageous to assay the activity of interest in the presence of other activities.

Finally, in this chapter, use of hplc itself as an aid in the purification of an enzyme activity is discussed. Its use is not restricted to the final stages of a purification. Its use of small sample volumes, its sensitivity, and its speed of separation make hplc an ideal analytical tool to monitor the efficiency of other steps and procedures that are used during a purification.

4.1 INTRODUCTION

4.1.1 The First Goal: Selection of the Biological Starting Point

Since enzymes are associated with living systems, the development of a strategy for preparation of an activity begins with the selection of a specific biological starting point. This selection may be difficult, since one can start with an organism like an elephant; an organ such as a liver; biological fluids such as blood or saliva; cells that occur naturally such as bacteria or protozoa; or, finally, cultured cells. Preparing an activity from an elephant will present problems quite different from those experienced when preparing that same enzyme from bacteria.

In order to deal with this choice, samples have been subdivided into three groups. In group I (see Fig. 4.1) are those samples that contain both a cellular and an extracellular compartment. The extracellular compartment can contain low molecular weight compounds such as the nutrients found in a fermentation broth as well as macromolecular materials such as collagen or proteoglycans found in tissues. Also included in this extracellular compartment are fluids such as tears, saliva, and urine.

While samples in group I contain cells, these usually will not be of the same phenotype. For example, tissue samples such as skin can contain several types of cells. Therefore, multicellular samples in group I must contain, as part of their purification scheme, a step for the separation of the different cell types.

Samples in group II are composed of homogeneous populations of cells, and samples in group III are formed by the lysis of these cells and can be anything from organelles to soluble purified proteins. It is often possible to start a "purification" scheme with a sample from group III. In this case the strategy for the purification would contain steps such as salt precipitation and chromatography.

4.1.2 The Second Goal: The Extent of the Purification or the End Point

One of the most important events in the history of biology was the demonstration of enzymatic activity. Next in importance came the isolation and purification of the activity and the demonstration that catalysis was

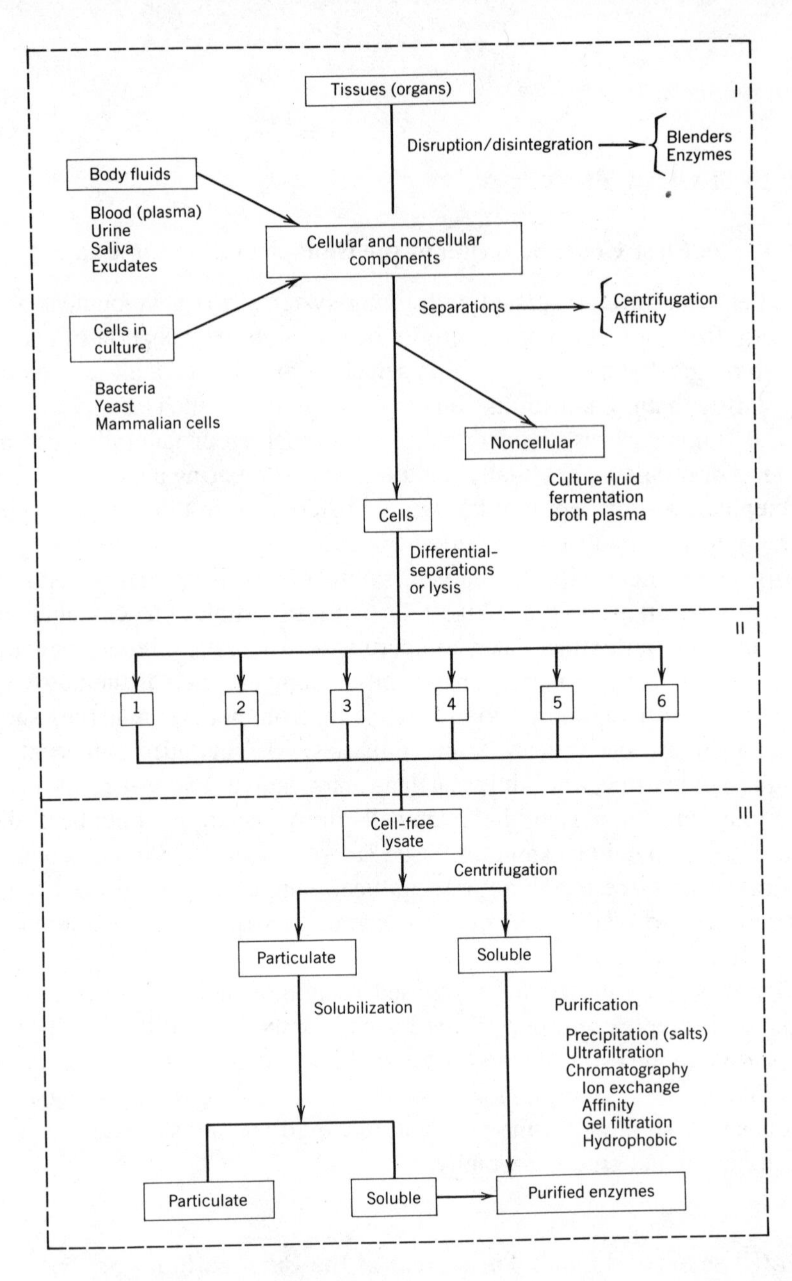

Figure 4.1 Anatomy of a purification scheme. Parts I, II, and III illustrate the different starting and ending points for purification. (I) The starting samples are composed of cells and extracellular materials and the goal of the purification is their separation. (II) The source of the enzyme is a heterogeneous population of cells, and the goal of purification is their separation into homogeneous populations (1–5). (III) The source of the enzyme is a cell, and the goal is the isolation of a subcellular organelle, fraction, or purified protein.

associated with protein molecules. Thus, advances in enzymology have been intertwined with advances in protein chemistry.

Enzyme isolation and purification as it is practiced today has progressed from the days when the tools available were few and limited to procedures such as "salting out" with ammonium sulfate, filtration, and centrifugation at rather unimpressive *g* forces. The procedures of the past involved a significant amount of art, as well as science, with the artistry demonstrated by the last step, which in most schemes usually involved crystallization of the enzyme. To watch a pioneer, like Moses Kunitz, coax enzyme crystals from an alcohol solution was inspirational. Unfortunately, videotapes were not around in those days to preserve such events.

Today, the enzymologist has a battery of techniques, including improvements in centrifugation and column chromatography, and enzymes can be purified to homogeneity faster than ever before. But should that be the goal of every purification?

Imagine looking under the hood of an automobile at its engine and trying to find, somewhere in the entire engine, a small screw 0.016 in. long with a 3/32 thread and a round head. One approach to finding such a screw, which is a minute part of the entire assembly, would be to take the engine apart as quickly as possible in order to obtain all the screws. These could then be spread out on a table top, and by careful inspection the screw of interest could be located.

Similarly, with an enzyme activity, if your goal is to isolate a special enzyme protein in order to characterize its size and shape and amino acid composition, then of course it would be wise to disrupt the starting material (organ, cell, or organelle) as quickly and completely as possible to obtain all the proteins, and then spread them out or fractionate them to locate the one of interest.

Returning again to the car engine, suppose we knew the screw was present, and now we wished to find its location within the engine. The strategy would have to be different. For example, it would be wiser to remove each part of the engine carefully and then disassemble each part in an attempt to locate the screw. In this way, when the screw was found, its location within the engine could be established.

When the question of the localization of an enzyme within the cell had to be answered, purification schemes were developed to take cells apart in a stepwise fashion. For this task a variety of tools were used, including rather commonplace scissors as well as sophisticated centrifugation and chromatographic techniques. And it was only after this careful dissection that an understanding emerged of where the enzymes were located.

But how do these enzymes function at the site where they are located? If the question were asked about the engine's screw, one answer might be found by examining that component of the engine to which the screw belonged. For example, if this screw was a part of the carburetor, it might function in regulating the intake of fuel or air, but it might operate in several other ways as well. In fact, it might not be possible to deduce its function merely by inspection, and to really find out how the screw worked, the intact carburetor would have to be returned to the engine and the engine started.

The same is true for an enzyme. Studies on what might be called its "interenzymatic" function would be best carried out while the enzyme was still a part of the organelle or complex to which it belonged. To date such studies have been difficult to perform because of problems in monitoring in a single assay the variety of enzymatic activities that can occur in such complexes. The introduction of hplc to assay enzymatic activity allows us to consider trying such interenzymatic functional studies, because with this method several activities can be measured simultaneously. Thus, one of the consequences of this advancement in methodology is that it is possible to measure the activity of an enzyme while it remains in a complex and through such studies deduce its function within a multienzyme complex.

A second consequence of this advancement in methodology is that it can affect the strategy employed for the purification of an enzyme, since the goal need not be to isolate the enzyme protein from all other enzyme proteins. Now the goal might be better stated: *Remove from the enzyme preparation only those components or structures that are not necessary for function.*

The scheme in Fig. 4.1 can be used to illustrate this point. While this scheme shows three separate starting points, it also shows potential end points. For example, if the sample is multicellular, the purification could be ended after the removal of the extracellular matrix or the medium or broth used to culture cells. Enzyme activities can be measured directly in the multicellular structures or in the extracellular compartment.

Alternatively, the purification can be continued, the cells lysed, and activities assayed after lysis. It is from this lysate that organelles such as nuclei and mitochondria are obtained. The disruption of organelles in turn produces soluble enzymes. Ultimately, the end point is determined by the individual study. If questions related to molecular weight, amino acid composition, or catalytic mechanism are to be answered, a pure protein will be required. In contrast, if questions relating one activity to another are of interest, then the purification should be ended at the lysate or organelle level.

In the following sections, the focus will be on samples from each of the three groups shown in Fig. 4.1. I leave my readers to answer the question: Where do I stop?

4.2 PREPARATION AND ASSAY OF ENZYMATIC ACTIVITIES IN SAMPLES OF TISSUES, ORGANS, AND BIOLOGICAL FLUIDS

4.2.1 Separation of Cellular from Extracellular Compartments

Samples Obtained Directly from an Organism. Tissues, such as connective tissue, and organs, such as skin or liver, can be thought of as being composed of at least two compartments: the cellular compartment and the extracellular compartment. Since enzymes can be localized in either compartment, one of the first problems is to separate the two compartments. Techniques should be used that will not damage the cells, since any damage is liable to cause leakage of the contents of the cellular compartment into the extracellular compartment. With tissues or organs, where the noncellular compartment is often a stable fibrillar matrix, a two-step procedure such as that shown schematically in Fig. 4.2 is helpful in separating the compartments. Often the matrix is disrupted by cutting or dicing with scissors, shearing in a blender, or grinding. However, with such physical techniques some cellular damage is unavoidable.

Next, the disruption of the matrix can be continued by treatment of the fragments with purified enzymes, which are often commercially available. These are ideal, since they can be chosen for their specificity and also chosen with the composition of the matrix in mind. In samples derived from mammalian tissue, the matrix usually contains collagen, and the enzyme collagenase can be used. Trypsin and other proteolytic activities have also been used with great success. As illustrated in Fig. 4.2, the

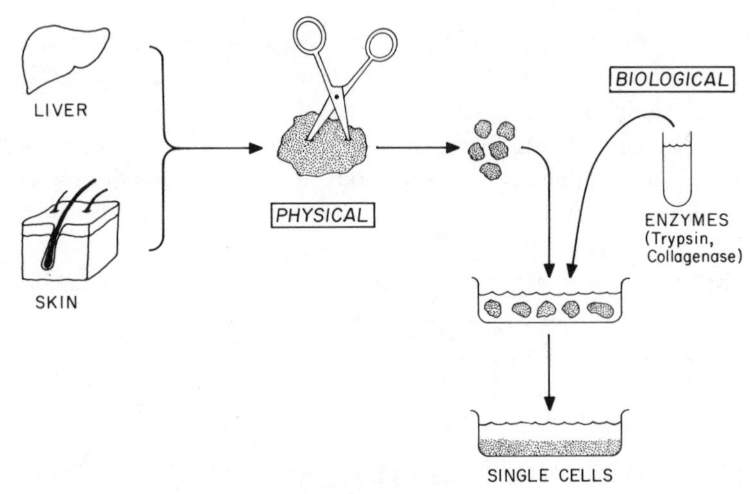

Figure 4.2 Methods used in the disruption of tissues, such as skin, or an organ, such as liver. The initial step is usually some physical technique; dicing with scissors is illustrated. The tissue fragments are then treated with an enzyme such as trypsin or collagenase to disrupt the fragments further to obtain single cells.

end result of this two-step procedure should be a solution containing intact cells, extracellular fluids, and extracellular components, including some insoluble fragments, some soluble components, and, of course, any enzymes added as reagents.

Samples Obtained from Tissue or Organ Culture. Animal tissues and organs can also be grown using a primary culture system by placing the sample on a support, such as an agar surface, a filter, or even a wire mesh, which can be positioned with the sample bathed in a solution of growth medium. A typical arrangement of the latter, shown in Fig. 4.3, consists of a culture dish containing a centrally placed well filled with culture fluid. Suspended over the well is a wire screen, which acts as a support for the tissue. After the well is filled with culture medium, the tissue is placed on the support and the dish is covered and incubated.

Samples obtained from such a culture system should be processed by the two-step procedure described above to obtain the individual cells free from the extracellular compartment. However, note that with cultured samples there is an additional extracellular compartment, which is the culture fluid used to support the growth of the sample. It is not at all un-

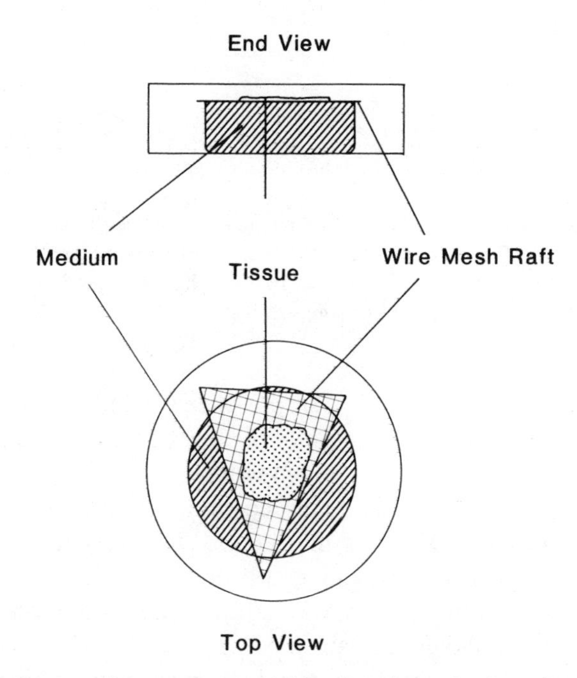

Figure 4.3 A diagram of the apparatus used for the culture of a tissue fragment or an organ. The dish contains a central well (diagonal lines), over which is placed a wire mesh raft to support the tissue. The well is filled with sufficient culture medium to make contact with the tissue. The dish is covered (not shown) and incubated under conditions used to maintain viability of the explant. (From Quintner *et al.*, 1982).

common to find enzymatic activities in this fluid. Some of these are normal constituents of the culture medium, while some are a consequence of growth of the sample.

Samples Obtained from Biological Fluids. As illustrated in Fig. 4.1, biological fluids have been placed in group I since they also are composed of two compartments. Such fluids include blood (often classified as a tissue), urine, semen, tears, and saliva. Many of these fluids contain cells as a normal component, while in others the cells represent a contamination. An example of the latter are those fluids that contact the "outside" such as saliva, which when collected can often contain microbes. Sometimes such microbes are the result of the collection procedure, and their numbers can often be controlled by careful technique. At other

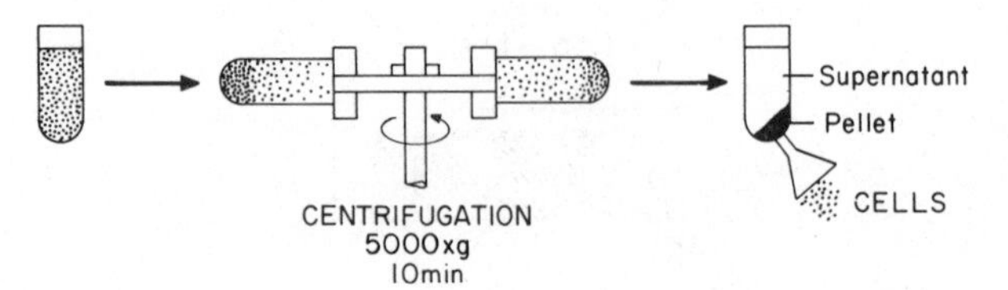

Figure 4.4 Harvesting cells from a culture medium. A sample of the culture medium is transferred to a centrifuge tube and subjected to centrifugation at rather low *g* forces such as 5000 *g* for a relatively short time, such as 10 min. The cells contained in the pellet can be recovered after the supernatant solution is decanted.

times, as is the case with urine, their presence can indicate an underlying disease process. In any case, the study of enzymes from such fluids again requires the separation of the two compartments.

Biological fluids, however, do not contain a fibrillar matrix material, and separation of the two does not require the two-step procedure described above. Often, centrifugation at a slow speed, such as 5000 *g* for 10 min as illustrated in Fig. 4.4, will suffice. The pellet produced during this centrifugation should contain most, if not all, of the cellular elements, including any microbes. It is advisable and informative to recover these pellets. A sample of the pellet and a sample of the supernatant solution should both be examined microscopically for their cellular content. The supernatant solution produced by the centrifugation can be assayed directly for enzymatic activities. However, if excess protein is present in the sample, it will have to be removed from the enzymatic assay samples before hplc analysis.

Samples Obtained from Cell Cultures. Cells of many types are now grown in liquid culture. These include not only mammalian cells but fungi, protozoa, and bacteria. In Fig. 4.1 these are placed together under the group I heading because in each case the noncellular elements in the fluid or fermentation broth should be separated from the cells before analysis. A low-speed centrifugation as shown in Fig. 4.4 should suffice. The supernatant fraction should be collected and assayed for the activity of interest, and the cells set aside for assay or lysis.

4.2.2 Assay of Activities in the Extracellular Compartment

Enzymatic activities will often be found in the extracellular fluid surrounding organs, tissues in biological fluids, or the medium supporting the growth of mammalian cells, bacteria, yeast, or fungi. The enzymatic composition of these fluids can vary considerably, and the assay of enzymatic activities in such fluids presents several major problems.

The first is the presence of proteolytic activities, which must be inhibited early in the procedure because their activity will result in the degradation of other enzyme proteins. Second, the amount of protein present in these fluids is usually in excess of what an hplc analytical column can handle without becoming clogged. And finally, these fluids often contain many low molecular weight compounds, either those added as nutrients or those present as a result of cellular metabolism. Since such compounds may resemble either the substrate or product, or both, of the enzymatic reaction under study, their presence in the reaction mixture could interfere with the assay. At the very least, such compounds will pass through the analytical column and appear on a chromatogram, confusing the experimental results.

A solution to the problem of low molecular weight compounds is to remove them, and a variety of methods are available, including dialysis and gel filtration chromatography. The removal of excess protein may be more complicated. It can be dealt with before the assay by further purification of the sample. Alternatively it can remain during the incubation and be removed after the assay by introducing a termination step to precipitate all proteins, which can then be removed by filtration. And finally, proteolytic activities can be eliminated by the addition of the "inhibitory cocktail" mentioned below.

A word of caution: Growth media that contain serum contain serum-associated enzymes. The presence of such endogenous activities must be considered in any purification scheme, since they can produce confusing results.

4.2.3 Assay of Activities in the Cellular Compartment

The purpose of the two-step procedure just discussed is the disruption of the samples in group I in order to separate the cells from the extracellular fluids. This procedure produces two samples: the solution of extracellular fluids and the cells. In all probability, however, the cells will not all be the same. In fact, depending on the complexity of the starting sample, the cellular population can be quite heterogeneous. Consider two organs, liver and skin, which are illustrated schematically in Fig. 4.2. Such structures contain several tissue types, including epithelia, connective tissues, and vascular and neural tissues. The preparation of enzymatic activities from such organ samples will require the separation of the different cell types from each other; otherwise, what during the course of a study might appear to be changes in enzymatic composition or activity will in fact be only a reflection of changes in composition of those cells making up the sample. Therefore, for any assay on a complex sample it is important to begin with a preparation of only one cell type. Many methods have been introduced to achieve this separation; some exploit differences in composition of the cells, others utilize differences in cell function, and still others take advantage of differences in composition at the cell surface.

For example, differences in composition of cells are often reflected in differences in density, and therefore cells can often be separated by centrifugation through a solution made of layers of different density—a technique called *buoyant density centrifugation*. At equilibrium the cells can be recovered from the solution at a position that balances their density. Such a procedure is illustrated schematically in Fig. 4.5. In this illustration, the starting tissue is represented as being composed of cells of three different types, where each type is denoted by a different symbol. A centrifuge test tube is filled with a series of sucrose solutions, each with a different density. In the example, three different solutions are used (a discontinuous gradient); the solution of highest density is, of course, placed on the bottom. Sucrose is often used for this purpose. If these cells are layered at the top of the solution, then after centrifugation each of the three cell types will be recovered in a different zone or band within the tube, its position dependent on its density.

Other methods also illustrated in Fig. 4.5 include the addition of chemicals (drugs) that by virtue of their specificity will produce lysis of one or two cell types, leaving a single cell type unaffected.

Two other techniques, both of which utilize antibodies raised against antigens specific to one of the cell types, have also been employed. In one of these techniques, antibodies can be attached to solid supports, which in turn can be used for column chromatography. In Fig. 4.5 the antibody specific for the round cells is shown attached to beads used to pack the column. When a solution of the three types of cells is passed through the column, the cells containing the antigen attach to the antibody and are retained, while the other, noninteracting cells pass directly through the column. In this way separation is achieved.

In a variation of the above technique, metallic iron, labeled Fe in Fig.

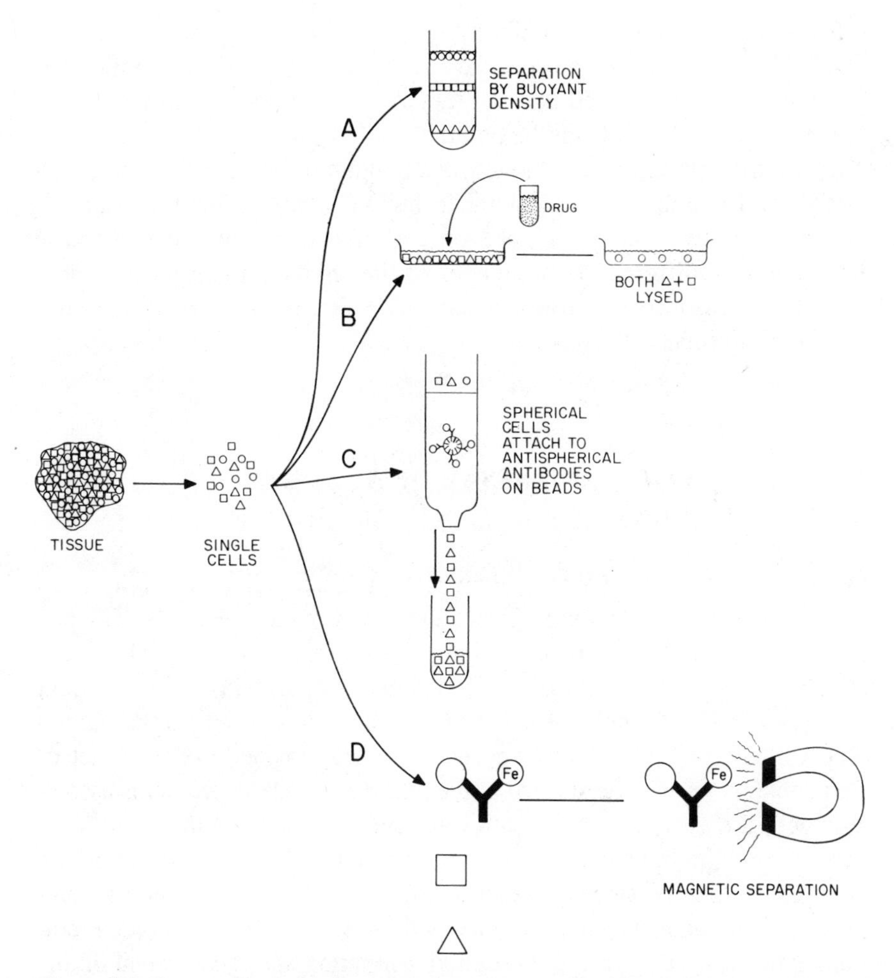

Figure 4.5 Some methods for the separation and isolation of homogeneous cell populations from tissue fragments. The tissue, composed of three cell types (○, □, Δ), is disrupted and a solution of single cells obtained. The cell types are shown being separated by (A) buoyant density centrifugation, (B) selective cellular lysis, (C) affinity chromatography, and (D) magnetoseparation.

4.5, can be attached to the antibody. The antibody is exposed to the cell, and after its attachment to a specific cell through interaction with the antigen the cells are exposed to a magnetic field, which will withdraw the "iron-containing" cell from the population.

Even after obtaining a homogeneous population of cells from an organ or tissue, it is important to remember that when comparing the results of enzyme activity studies obtained on cells from one animal with those from another, differences in the ages of the animals, their genetic backgrounds, or even their nutritional status can alter the enzymatic activities recovered in their cells.

4.3 PREPARATION AND ASSAY OF ACTIVITIES IN INTACT CELLS

In group I, we have considered the question of obtaining an activity from a tissue or organ, from a biological fluid, or from cultured cells. The primary task in all these samples was the separation of the extracellular and cellular compartments. Next, the problem of separation of the different cell types within the cellular compartment was considered. In the section that follows, the cell will be opened for a look inside. However, let us first consider briefly the surface of the intact cell and the problems associated with the assay of any activities that might be located there.

Setting up an hplc assay for activities on intact cells requires no major changes except that since the reaction mixture will contain cells, the samples for hplc analysis cannot be injected directly onto the analytical column. Thus, the reactions will require termination and the removal of any precipitated proteins (see Chapter Three). Termination can be accomplished by centrifugation at a low speed or by filtration (see Fig. 3.9). However, care should be taken to avoid any cell breakage, particularly if the product of the primary or even secondary reactions can also be found as a naturally occurring intracellular component.

Consider an assay for a cell-surface ATPase where the reaction product is ADP. A reaction mixture is prepared that contains ATP as the substrate. The addition of the cells will start the reaction, and ADP will be produced directly into the extracellular solution. Lysis of the cell during this assay will release cellular ADP into the incubation medium, altering the results of the assay.

4.4 PREPARATION AND ASSAY OF ACTIVITIES IN SUBCELLULAR SAMPLES

The lysis of cells has its consequences. For example, since most lytic procedures involve disrupting both the plasma membrane and the membranes surrounding internal organelles, opening a cell will result in a loss of boundaries that would otherwise segregate enzymes from degradation. Therefore, following the disruption of most cells, proteolytic activities from some locations such as lysosomes gain access to areas from which they normally would have been excluded. Thus, precautions must be taken if problems arising from proteolytic activities are to be kept at a minimum or, better yet, prevented.

In the absence of any specific information about the nature of such activities, it is often best to mix a cocktail containing several or all of the available proteolytic inhibitors. These include such compounds as 1,10-phenanthroline, soybean trypsin inhibitor, leupeptin, benzamidine, antipain, aprotinin, phenylmethanesulfanyl fluoride, and diisopropylfluorophosphate. As illustrated in Fig. 4.6, such a cocktail is best added to the buffer in which the cellular lysis will be carried out as a precaution against any enzyme destruction.

Opening the cells can have other consequences. For example, many enzymes require cofactors for catalytic activity. If, as a result of the lysis of cells and organelles, the enzymes become separated from these cofactors and a loss of activity can result. Also, since the activity of most enzymes is concentration-dependent and since lysis usually results in the dilution of intracellular components, the activity of enzymes present at low intracellular concentrations can be lost. Both these problems are difficult to guard against.

In addition to these more general concerns are those questions concerning the localization of an enzyme activity. The location of an enzyme can determine the type of cell lysis, since it could be more advantageous to lyse the cell completely or in such a manner that the organelles are left intact. For example, some lysis methods such as sonication completely dis-

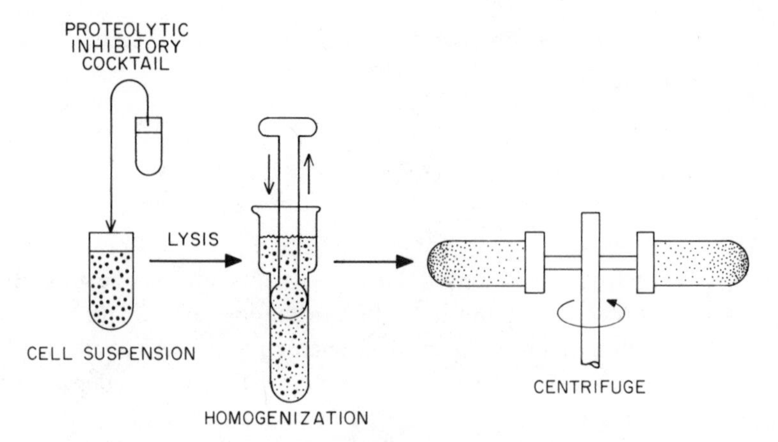

Figure 4.6 Overview of steps involved in the preparation of a cell-free lysate. The cells are resuspended in a buffered solution at a specified cell density. To this suspension is added a "cocktail" containing several proteolytic inhibitors. The cells in the suspension are lysed (in the illustration a homogenization is shown). Finally the lysate is subjected to a very low speed centrifugation such as 5000 *g* for 10 min to remove unbroken cells.

rupt mitochondria, nuclei, and Golgi systems. If an activity is localized in an organelle such as a mitochondrion, it would seem sensible to adopt a method that leaves these structures intact in order to facilitate their separation from the rest of the cellular debris. Thus, for the isolation of mitochondrial enzymes, sonication cannot be considered the method of choice for cell lysis.

Those methods commonly used for lysis include such techniques as the French press, the sonicator, and the blender or homogenizer. In Fig. 4.6, a homogenizer is represented in sequence with the other steps involved in the lysis process. The choice of the lysis method, while dependent on the localization of the enzyme, is also a function of the type of cell under consideration. For example, when lysing bacterial cells with rigid cell walls, the French press may be one of the few physical methods that works. Alternatively, for disrupting cells that have fragile cell membranes, homogenization as shown in Fig. 4.6 is often adequate.

Regardless of the procedure that is finally adopted, it would be wise to measure its success. One of the easiest techniques for accomplishing this assessment is a microscopic examination of samples taken from the lysate after each step of the procedure. Such information is very useful, particularly when experimenting with either a new type of cell or a new method for lysis.

4.5 INITIAL PURIFICATION AND ASSAY OF ACTIVITIES IN CELL-FREE LYSATES

While it is possible and often necessary to assay enzymatic activities directly in the lysate, it is often helpful to remove any remaining unbroken cells and any other nonbiological debris such as sand grains or glass beads that might have been added to facilitate the breakage. Again, centrifugation at a low speed, such as 5000 rpm, for about 10 min should be sufficient.

The supernatant fraction obtained in this step can be centrifuged at 30,000 *g* to produce a second supernatant fraction often referred to as an S-30 fraction. Centrifiguation of this S-30 fraction at 100,000 *g* will produce a third supernatant fraction called an S-100 fraction. Each centrifugation removes insoluble material of decreasing size or mass. For example, while the pellet from the S-30 fraction contains many large organelles, such as intact mitochondria, the pellet from the S-100 fraction contains many smaller structures such as ribosomes, the endoplasmic reticulum, and other membranous structures.

Many enzymes are soluble and will be recovered in the S-100 fraction. In addition, their activity can be measured rather conveniently by adding a sample of the S-100 fraction directly to a reaction mixture. Of course, since the S-100 fraction will contain an excessive amount of extraneous protein, it will be necessary to terminate the reaction and filter the sample prior to injecting it onto the hplc for analysis.

Note that while the S-100 fraction can be used in this form, it would be best to have it dialyzed to remove unwanted low molecular weight compounds prior to its use in an assay. In addition to dialysis, some extraneous material can be removed by a simple salting out of some of the proteins. Ammonium sulfate is often added for this purpose and can be used to remove unwanted proteins or to precipitate the enzyme in question. The ammonium sulfate should be removed prior to using the sample in the assay, because the salt might affect activity. Again, dialysis can be used, or alternatively the sample can be passed through a gel filtration (G-25) column.

If it is necessary to purify the enzyme further, the next step should be one that has a high capacity, that is, one that can process large amounts of protein. Such steps, however, often are not very specific. For example, salting out is a technique of high capacity but low selectivity. This step should be followed by steps of decreasing capacity but greater selectivity. Such techniques include ion-exchange, gel filtration, or affinity chromatography and even hplc itself. The choice should take into account what is known about the protein, its size and shape, its solubility, and even its substrate specificity. Also, the quantity of protein required should be considered. If only analytical amounts of the enzyme are needed, then several methods including hplc can be included.

4.6 HPLC FOR PURIFICATION OF ENZYMES: A BRIEF BACKGROUND

Early separations of proteins by hplc were made with controlled pore glass beads with a coating of 1% polyethlene glycol. The technique was later modified with a 3% polyethylene glycol coating on the beads and used for the separation of plasma proteins. The development of noncompressible ion-exchange supports has allowed various laboratories to separate the isoenzymes of lactate dehydrogenase. Refinement of this anion-exchange support to a microparticulate size (40 μm) has enhanced the resolution of lactate dehydrogenase and creatine kinase isoenzymes as well as the purification of alkaline phosphatase.

Reverse-phase chromatography has also been used to separate proteins. However, the required use of alcoholic gradients or paired-ion reagents with the reverse-phase support should be avoided, as they could cause inactivation of the enzymatic activities.

High pressure gel permeation chromatography (hpgpc) has been developed to correspond to gel filtration where large molecules are partially excluded from the porous coating of the support and thus are eluted before smaller molecules. A silica-based packing with an organochlorosilane chain containing other functional groups has been used to separate plasma proteins on the basis of their molecular size as well as by their ionic, polar, and hydrophobic interactions with the column packing and the mobile phase.

In addition to its use for the purification of an activity, there are several places in any overall purification scheme where hplc is useful as an analytical tool. For example, hplc can be useful in the establishment of gradients. Instruments have been manufactured for use with hplc that can control the flow and mixing of solvents and thereby generate gradients with a variety of concentrations and "shapes." In addition, since it is possible to carry out separations on the hplc column in a comparatively short time, a number of these gradients can be applied to an analytical scale column, and the one best suited to the separation established fairly rapidly. Armed with this information it is a relatively simple matter to carry over these gradient conditions to a non-hplc ion-exchange column.

The speed with which it is possible to perform an analytical run makes hplc useful for what might be called a pseudo-preparative function. For example, if rather modest amounts of a purified protein are required—for example, to carry out some analysis by polyacrylamide gel electrophoresis or even for antibody production—and if the separation and purification obtained on the hplc column is adequate, then it is often possible to produce enough material for such purposes by merely repeating the same run on a analytical column several times. By collecting the appropriate fraction it is possible to generate sufficient material to advance to the next stage of the purification or perform the experiments of interest. Again, it is the speed of the separation that makes this approach feasible.

Finally, hplc can also be used as an analytical method to monitor the efficiency of different purification methods. For example, imagine that gel filtration chromatography has just been carried out on a sample of an S-100 fraction. Several peaks are observed. A question usually asked at this point relates to the homogeneity of each of the peaks. HPLC is ideal to answer this question since each of the peaks can be analyzed by hplc in just a few minutes. Also, different columns can be used for the analysis, and therefore the homogeneity of the peak can be verified under a variety of conditions.

As another example, imagine that an S-100 fraction has been prepared, ammonium sulfate has been added, and the fraction that precipitates between 30 and 50% has been obtained. This sample is subjected to affinity and ion-exchange chromatography. Have these procedures been successful in removing extraneous proteins? While it is possible to answer this question by a determination of total protein content, it could be more informative if each of the samples could be analyzed for its constituent proteins.

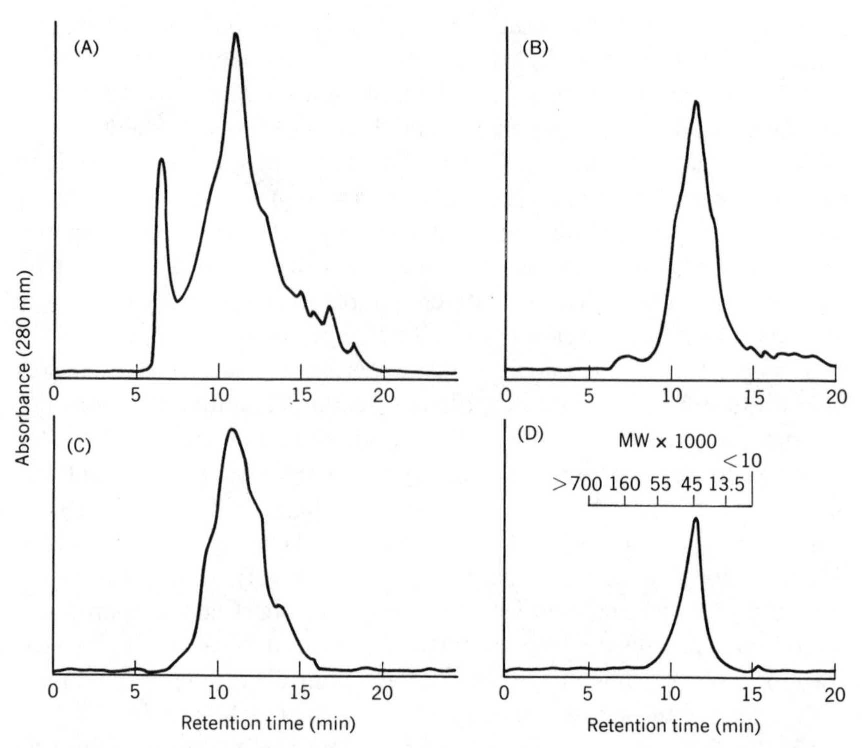

Figure 4.7 The use of hplc to monitor enzyme purification. Profiles A–D obtained by gel-filtration chromatography during the purification of the enzyme sAMP synthetase. The column was a TSK-250 (BioSil, 7.5 mm × 30 cm), and the mobile phase was 0.1 *M* potassium phosphate (pH 6.0). The column was monitored at 280 nm. Profiles obtained after (A) 30–50% ammonium sulfate precipitation, (B) affinity chromatography, (C) ion-exchange chromatography on DE-52, and (D) hplc ion-exchange chromatography on AX-300.

The speed of the hplc analysis makes such a determination possible. Figure 4.7 shows four panels, representing the analysis of samples after each of four stages of a representative purification. The hplc analysis was made using gel filtration hplc, and as shown this could be accomplished in 20 min.

A comparison of the protein profiles obtained reveals that proteins present in the sample prepared by ammonium sulfate fractionation (Fig. 4.7A) were removed following affinity chromatography (Fig. 4.7B) and ion-exchange chromatography (Fig. 4.7C). The profile of the activity following ion-exchange hplc (Fig. 4.7D) shows considerably less protein than was originally present.

4.7 STRATEGY FOR USE OF HPLC IN THE PURIFICATION OF ACTIVITIES

Hplc has received a great deal of attention as a method for the purification of enzymes, and the results of such studies are rapidly pervading the literature. The reader should consult recent reviews for information concerning the enzyme of interest.

Consistent with the purpose of this volume, the general principles of how to begin the purification procedure using hplc will be presented. Over the years, the experience that has been gained in the purification of enzymes has led to the general approach for the purification presented in this section.

In order to obtain the separation needed for purification, the first few runs are carried out to verify solubility and to get some idea of the complexity of the sample. Separation of the proteins can require modification of several variables, including the concentration range of the salt used for the gradient as well as the salt itself. For example, as shown in Fig. 4.8A,B the range of salt concentration used in the gradient can have a significant effect on the elution profile of a series of proteins. The pH might also be changed, and Fig. 4.8C,D show the effect of this pH change on the separation. During this initial phase, when the purpose is merely to establish conditions for the separation, it is usually not necessary to collect fractions or assay enzymatic activity. Throughout this phase of the work, the separations can be monitored at 280 or 230 nm in the absence of aromatic amino acids in the protein.

When conditions for the separation have been attained, the enzyme should be located. Another run should be carried out for this purpose, the fractions collected, and the activity of each fraction determined. Figure 4.9 illustrates the results of such an analysis. The absorbance profile obtained at 280 nm of the sample as fractionated on an ion-exchange column is shown. In Fig. 4.9A the fractions that were collected were assayed for three different enzymes; the results are shown in Fig. 4.9B,C,D. Three activities are present, and each activity has been separated into two components. Additional runs may be necessary to obtain

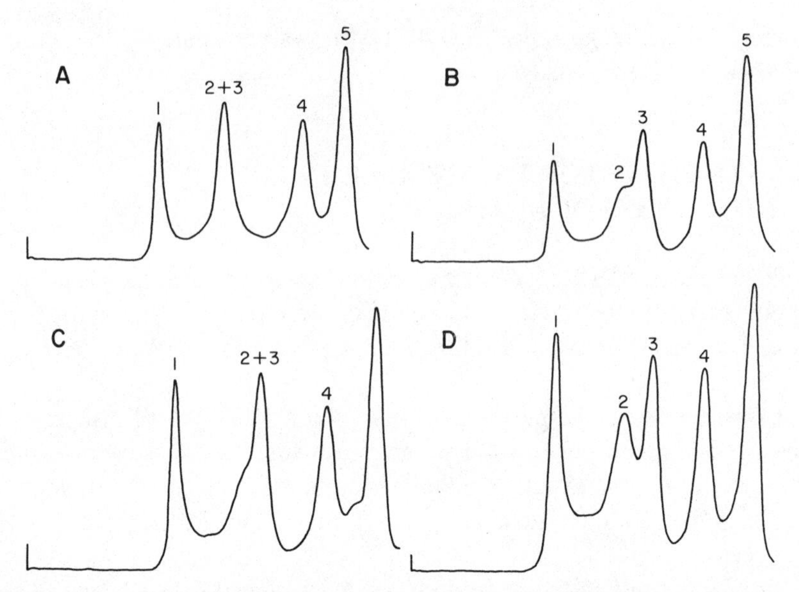

Figure 4.8 Effect of ionic strength and pH on elution profile of proteins on gel-filtration hplc. Proteins include (1) thyroglobulin, 670,000; (2) gamma globulin, 158,000; (3) ovalbumin, 44,000; (4) myoglobulin, 17,000; and (5) vitamin D-12, 1350. The column was eluted with sodium acetate (pH 6.8), (A) at 50 m*M* and (B) at 100 m*M*. The column was also eluted with (C) 100 m*M* sodium acetate (pH 6.0) and (D) 150 m*M* sodium acetate (pH 6.0).

sufficient material for subsequent purification. If an hplc step is introduced early in the purification, several runs will be required to obtain enough sample for additional purifications.

To illustrate the rapidity of hplc, particularly in comparison with the more conventional techniques, the separation of the same sample has been carried out by conventional ion-exchange chromatography. Figure 4.10 compares the two procedures. These data show that where 14 h was required for the traditional method, only about 45 min is required with hplc. Therefore, the total time needed to carry out this purification, not counting the time for the enzyme assay, could be as short as 3–4 h. If necessary, the chromatography step could be completely automated. Finally, since each run will use only a fraction of the total volume of the starting material, the entire procedure will be economical.

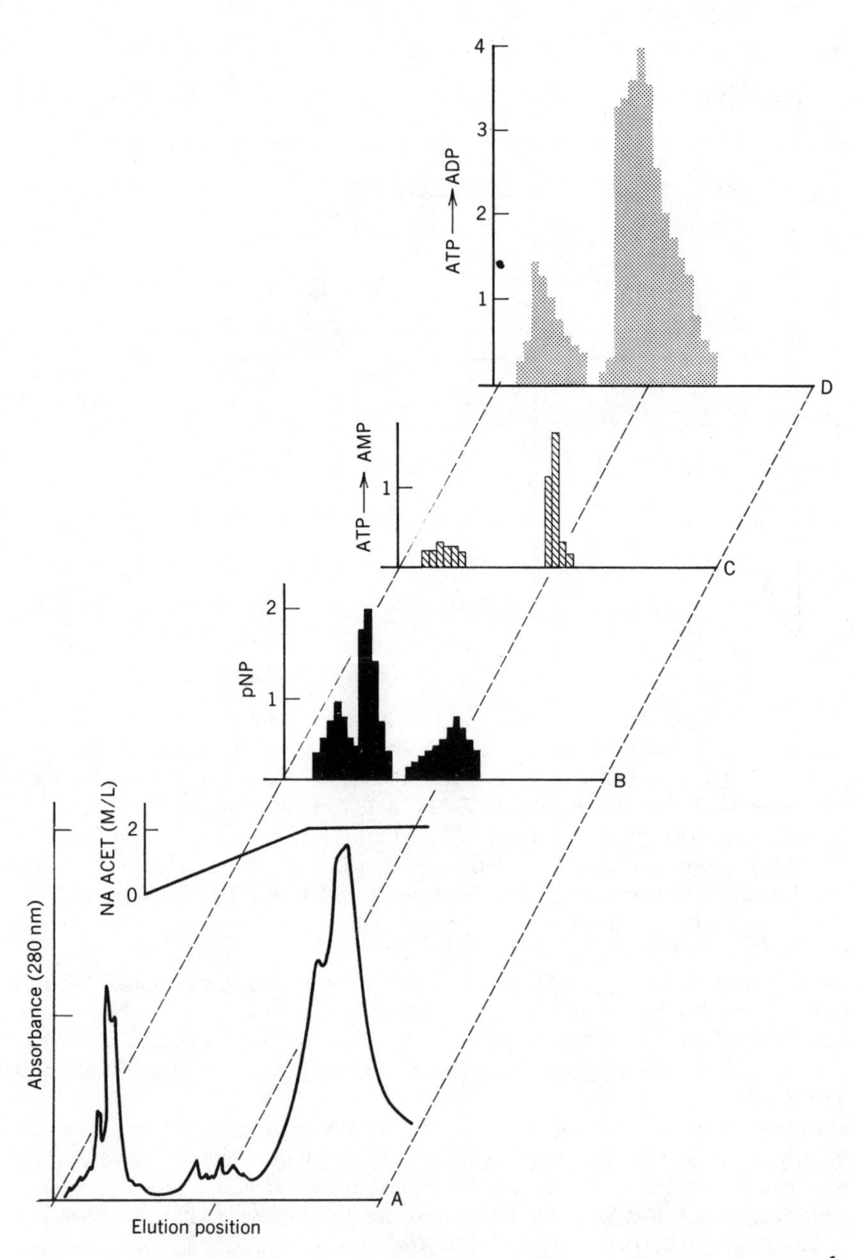

Figure 4.9 Enzymatic activities of fractions following hplc chromatography. A partially purified preparation was fractionated by ion-exchange hplc (AX-300) with a mobile phase of 0:1 *M* potassium phosphate. Proteins were eluted with a gradient of sodium acetate. Column eluent was monitored at 280 nm. Fractions were collected, and each fraction was assayed for three different activities.

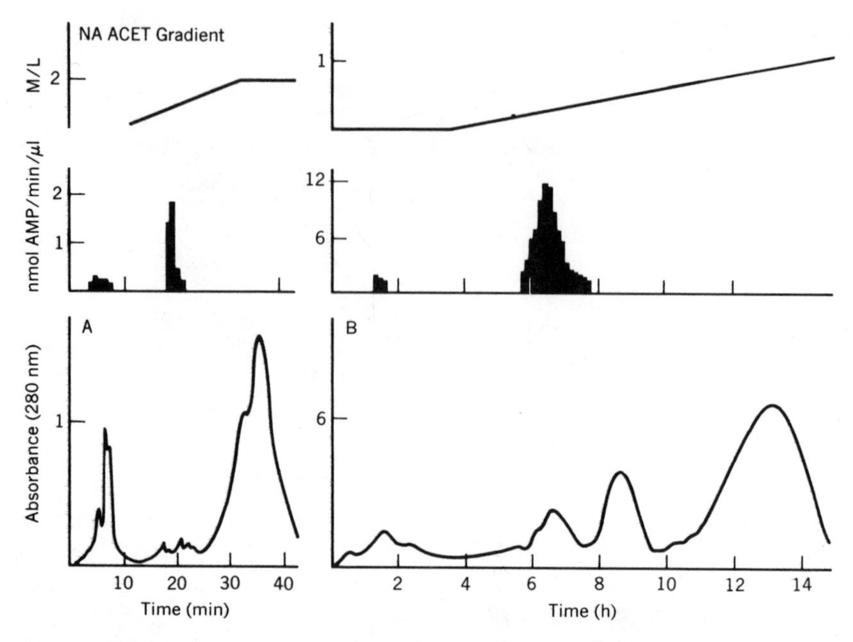

Figure 4.10 Ion-exchange chromatography of a detergent-solubilized membrane fraction. (A) Approximately 2 mg protein from the fraction was injected onto an AX-300 anion-exchange column (250 × 4.1 mm) that was equilibrated with a 20 m*M* sodium acetate buffer (pH 6.3) containing 0.1 m*M* Z-314. A 20 min linear salt gradient of 20 m*M* to 2 *M* sodium acetate (pH 6.3) with 0.1 m*M* Z-314 (top panel) was used to elute proteins. The protein profile is shown in the bottom panel. The flow rate was 1 mL/min, and the absorbance was monitored at 280 nm. Fractions of 1 mL were collected and immediately put on ice. For the amount of ATP pyrophosphohydrolase activity contained in each column fraction, the following assay was performed. To 75 μL of each column fraction was added a reaction mixture such that 0.4 m*M* ATP, 0.4 m*M* $MnCl_2$, and 50 m*M* Tris-HCl (pH 7.4) were in 100 μL final volume. The reactions were run at 31°C for 2 h and terminated at 155°C for 1 min. An analysis of the reaction components was done by reverse-phase hplc using a μBondapak C-18 column. The mobile phase was 65 m*M* KH_2PO_4 (pH 3.6), 1 m*M* tetrabutylammonium phosphate, and 2% acetonitrile. 20 μL samples of the assay tubes were analyzed, and the amount of AMP formed by the pyrophosphohydrolase in each column fraction is expressed as nmol AMP/min per microliter. (B) Approximately 60 mg of protein from a detergent-solubilized membrane preparation was applied to a DE-52 column (30 × 2.5 cm) that had been previously equilibrated with a 20 m*M* sodium acetate buffer (pH 6.3) with 0.1 m*M* Z-314. After washing with 200 mL of buffer, a linear salt gradient (top panel) of 20 m*M* to 2 *M* sodium acetate in 2 L (total volume) was used to elute the proteins. Fractions of 15 mL were collected at 4°C, and absorbance was monitored at 280 nm. The protein profile is shown in the bottom panel. For the amount of ATP pyrophosphohydrolase activity contained in each column fraction, the enzymatic assay as described above was performed.

4.8 PROBLEMS RELATED TO THE ASSAY OF ACTIVITIES FOLLOWING THEIR PURIFICATION BY HPLC

While most of the problems in the assay of an activity purified by hplc are expected and usually experienced in chromatographic work with enzymes, the introduction of this technique into the purification scheme may introduce some problems if the fractions obtained from the hplc purification step are to be measured for enzymatic activity. For example, the salt in each fraction may inhibit any enzymatic activities it contains. And since, when ion-exchange hplc is used, the salt concentration will vary in the fractions, it is prudent to study the effects of salt, at the concentration used for elution, on enzyme activity before the chromatography. If found to be detrimental, the salt will have to be eliminated or at least reduced in concentration before the chromatography. Since the inactivation of enzyme activities is not always reversible, removing the salt by dialysis may not be a successful way to overcome its effect.

In addition, the detergents often added to enhance the solubility of proteins can cause problems, since they can inhibit activities directly or their removal can result in the loss of solubility and thus cause precipitation. It should be noted, however, that when hplc is used for the purification, more often than not the protein concentration of the sample will be low, and the concentration of protein in each of the fractions will be lower still. Therefore, the formation of a precipitate may not be visible, and monitoring the sample at 320 nm, a wavelength useful in monitoring any light scattering, may be the only reliable method to detect precipitates. As a precaution it is good practice to dialyze the sample against the mobile phase that is to be used for the elution. When a gradient is used, dialysis against the starting and ending buffer should be carried out as well.

Many detergents, such as Triton X-100, absorb in the ultraviolet range and will therefore interfere with the detection of proteins at 280 nm. The use of nonabsorbing detergents will eliminate this problem. Detergents can also interfere with the operation of ion-exchange columns, and a reduction in detergent concentration may be required for the correct performance of the ion-exchange packing materials.

As an example, we found that a 10% zwitterion detergent was required for complete solubilization of the membrane fraction. However, we also found that the presence of detergent blocked the retention of some proteins on the ion-exchange column and, further, that dialysis of the proteins to remove the detergent resulted in the prompt precipitation of the protein. The problem was solved by trial and error. By dialyzing the protein against detergent solutions of various concentrations, a concentration was found low enough to permit ion-exchange chromatography but high enough for solubility.

Again, it should be mentioned that because salt solutions often act to precipitate detergents, as in the precipitation of sodium dodecyl sulfate by potassium, it is necessary to check the solubility of the protein solutions in the detergent against the salt solution at the concentration that will be present at the conclusion of the gradient.

Two other problems often arise following the use of hplc for purification. The first has to do with the volume of the sample used for assaying activity. Having successfully completed the ion-exchange step, it will be necessary to determine enzymatic activity. These determinations will be performed on samples taken from a series collected during the course of the purification. However, with hplc purification the volume of each sample collected will probably be no more than a few hundred microliters and often less. Further, there will usually be only a small number of samples. This situation in hplc is in contrast to that found in traditional chomatography, where the volume of each sample can be in the milliliter range and the total number of samples or fractions collected can be in the hundreds.

Since, in order to locate the enzyme, it is necessary to assay each of the fractions for activity, the concentration of salts will not be the same in each fraction when gradients are used. And since salts often inhibit activity, false data may be obtained on the distribution of activity across the column if salts remain in the sample.

Finally, the pH at which the purification is carried out may not be suitable for the assay. To solve this type of problem, many investigators, particularly those more familiar with fractions containing volumes such as 10–20 mL, usually adjust the sample solution to the assay conditions either by dilution of an aliquot of the fraction into the buffer used for the enzyme assay or by removing 1–2 mL from each and, by dialysis, changing the buffers.

Of course, with the small volumes involved, dialysis is often out of the question. Dilution will also usually consume most of a given fraction.

Therefore, it is necessary to be prepared to carry out the separation step more than once.

Most if not all the problems associated with salt, pH, the presence of organics, can be minimized if not eliminated following concentration using newly developed microconcentration systems.

4.9 SUMMARY AND CONCLUSIONS

Two considerations dominate the development of a strategy for the purification of enzymatic activities: the choice of the sample to be used as the source of the enzyme and the extent of the purification.

The samples that can be used as a source of starting material can be divided into three groups: multicellular (I), cellular (II), and subcellular (III).

For those samples in group I, which includes tissues, organs, biological fluids, and cultured cells, one task is separation of the cellular from the extracellular compartment and another is obtaining a homogeneous population of single cells. These cells constitute the samples of group II. Group III contains those samples obtained after cell lysis and includes organelles and purified proteins.

Hplc can be useful for the purification of proteins to homogeneity. It can also be useful as an analytical tool to monitor the purification of proteins using other more preparative procedures.

When hplc is used to purify enzymes and the enzymes must be located by an analysis of the fractions collected during the separation, problems may develop due to the solvents used in the purification.

GENERAL REFERENCES

Cell Separation Techniques

Owen, C. S., Magnetic cell sorting, in T. G. Pretlow and T. P. Pretlow, eds., *Cell Separation: Methods and Applications,* Vol. 2, Academic, Orlando, FL, 1982.

Pretlow, T. G., and Pretlow, T. P., Sedimentation of cells: An overview and discussion of artifacts, in T. G. Pretlow and T. P. Pretlow, eds., *Cell Separation: Methods and Applications*, Vol. 1, Academic, New York, 1981.

Regnier, F. E., HPLC of membrane proteins, in J. C. Venter and L. C. Harrison, eds., *Receptor Purification Procedures,* Vol. 2, Liss, New York, 1984.

Waymouth, C., Methods for obtaining cells in suspension from animal tissues, in T. G. Pretlow and T. P. Pretlow, eds., *Cell Separation: Methods and Applications,* Vol. 1, Academic, New York, 1981.

Quintner, M. I., Kollar, E. J. and Rossomando, E. F. *Expl. Cell Biol.* **50**:222 (1982).

Protein and Enzyme Purification by HPLC

Hearn, M. T. W., Reversed-phase high performance liquid chromatography, in W. B. Jakoby, ed., *Methods in Enzymology,* Vol. 104, Academic, Orlando, FL, 1984.

Regnier, F. E., High performance ion-exchange chromatography, in W. B. Jakoby, ed., *Methods in Enzymology,* Vol. 104, Academic, Orlando, FL, 1984.

Schmuck, M. N., Gooding, K. M., and Gooding, D. L., *J. Liq. Chromatogr.* **7**:2863 (1984).

Unger, K., High performance size-exclusion chromatography, in W. B. Jakoby, *Methods in Enzymology,* Vol. 104, Academic, Orlando, 1984.

Chapter Five

Survey of Enzymatic Activities Assayed by the HPLC Method

Overview

In this chapter those enzymatic activities which have been assayed by the hplc method are surveyed. These include enzymes involved in the degradation of proteins and the metabolism of catecholamines, amino acids, polyamines, heme, carbohydrates, steroids, purines, pterins, and polycyclic hydrocarbons.

Each assay will be presented according to the scheme used throughout this book. The primary reaction will be introduced, followed by the methods used for separation, including stationary phase, mobile phase, and the method of elution. The detection technique will be presented along with any considerations for calibration or quantitation.

The reaction mixture, including buffers and pH, the sequence for initiating the reaction, and the process used for termination will be described. Next the methods used in the preparation of the sample including centrifugation, filtration, or any other type of purification before the analysis by hplc will be presented. Finally, the source of the enzyme activity will be reviewed, and the steps used in the purification will be described briefly.

The specific reference from which the description of the enzyme assay was taken is cited in the text. The reference section has been organized into subdivisions that correspond to those used throughout the chapter.

The general reference section contains additional citations which are relevant to the enzymes described, to provide readers with a more extensive survey of the hplc assay methods that have been developed for these activities.

5.1 INTRODUCTION

From what has been presented in the previous chapters, it should be clear that in order for the hplc method to be used to assay an enzymatic activity, a stationary support phase and mobile phase must both be available for the separation of the substrate and the product.

Of course, the converse is also true. If a system for separation of two compounds that have a substrate–product relationship has been developed, it may be used as the basis for developing an enzymatic assay.

5.2 CATECHOLAMINE METABOLISM

5.2.1 Tyrosine Hydroxylase (Haavik and Flatmark, 1980)

Tryosine hydroxylase is a monooxygenase that catalyzes the conversion of L-tyrosine to dopa. The activity is found in peripheral and cholinergic neurons and chromaffin cells of the adrenal medulla. Two hplc methods have been developed for the assay of this activity.

In one method, the dopa formed during the reaction was partially purified by ion-exchange and aluminum oxide chromatography and the amount present quantified by reverse-phase hplc (ODS column). The mobile phase consisted of 0.1 *M* potassium phosphate buffer at pH 3.5. The column was eluted isocratically and the eluent monitored using an electrochemical detector.

The volume of the reaction mixture used with the hplc assay was only about one-fifth the volume usually required with other assay methods, resulting in a considerable saving in reagents. The reaction was terminated with perchloric acid, the pH of the solution was returned to about 8 with potassium carbonate, and the sample was clarified by centrifugation. First the supernatant solution was purified using the double-column chromato-

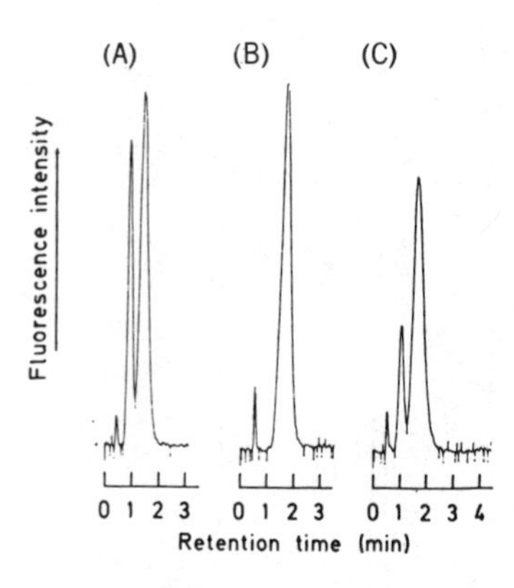

Figure 5.1 (A) Chromatogram of 384 pmol of dopa (1.03 min) and 1.92 nmol of L-tyrosine (1.55 min). (B), (C) Chromatograms of an acidic ethanol extract of an incubation mixture in the assay of tyrosine 3-monooxygenase activity of bovine adrenal medulla microsomes. (B) Zero-time control with a single peak of L-tyrosine (1.55 min). (C) The formation of dopa (1.03 min) following a reaction period of 20 min. Volumes of 20 μL of the diluted (twice) incubation mixture were injected into the liquid chromatograph: $\lambda_{ex} = 281$ nm, and $\lambda_{em} = 314$ nm. (From Haavik and Flatmark, 1980.)

graphic procedure mentioned above, and then the samples were injected onto the hplc column and analyzed for dopa.

With the electrochemical detector, it is possible to quantitate the amount of dopa present. The presence of pterins, cofactors required for activity, can also be detected.

Also, as noted, the hplc method eliminates the need to know the concentration of tyrosine in the tissue. Such information would be required when activities are assayed with radiochemical methods, since any unlabeled substrate, in this case tyrosine, would reduce the specific activity of the labeled tyrosine.

In the second method, there was no prepurification: dopa was measured directly. The separation of dopa and tyrosine was carried out on a cationic ion-exchange hplc (sulfonated fluorocarbon polymer) protected with a precolumn packed with silica pellets. The column was eluted with a mobile phase of 10 m*M* acetate buffer (pH 3.70) with 1% (v/v) propanol. The separation of tyrosine and dopa is shown in Fig. 5.1A. Because of its unique natural fluorescence, the dopa can be monitored with a fluorescence detector without interference from endogenous substances. An excitation wavelength of 281 nm was used with the emission at 314 nm.

The reaction mixture contained the substrate L-tyrosine. Benzyloxyamine was added to inhibit any secondary reactions catalyzed by the enzyme aromatic L-amino acid decarboxylase, an activity often present in these samples. Chromatograms of samples taken during an incubation are shown in Fig. 5.1B, a zero time control with a single peak of tyrosine,

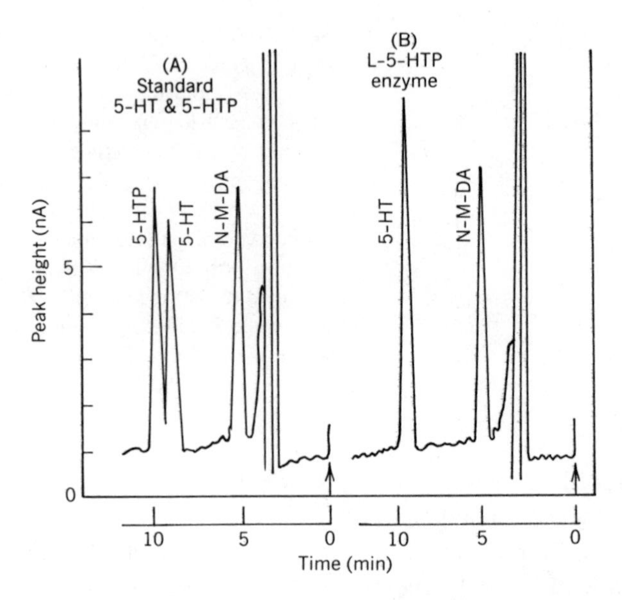

Figure 5.2 Hplc elution pattern of L-5-HTP decarboxylase incubation mixtures with homogenate of rat cerebral cortex as enzyme. The standard incubation mixture contained 5 mg of rat cerebral cortex. (A) Standard mixture of 50 μL containing 17.5 pmol each of L-5-HTP, 5-HT, and *N*-methyldopamine (*N*-M-DA). (B) Experimental incubation with L-5-HTP; 250 pmol of *N*-M-DA was added to each sample after incubation. (From Rahman et al., 1980.)

and Fig. 5.1C, after 20 min of incubation showing the peak of dopa formed as a result of enzymatic activity.

The samples, prepared from bovine adrenals, were homogenized, and the homogenate was purified further by centrifugation in 0.2 *M* sucrose.

5.2.2 5-Hydroxytryptophan Decarboxylase (Rahman et al., 1980)

Aromatic L-amino acid decarboxylase catalyzes the decarboxylation of L-5-hydroxytryptophan (L-5-HTP) to serotonin (5-HT). In the assay, L-5-HTP was used as the substrate and the formation of 5-HT was measured.

The separation of reactant from product was carried out on a reverse-phase column eluted isocratically with an elution buffer of 0.1 *M* potassium phosphate containing 10% methanol at pH 3.2. *N*-Methyldopamine

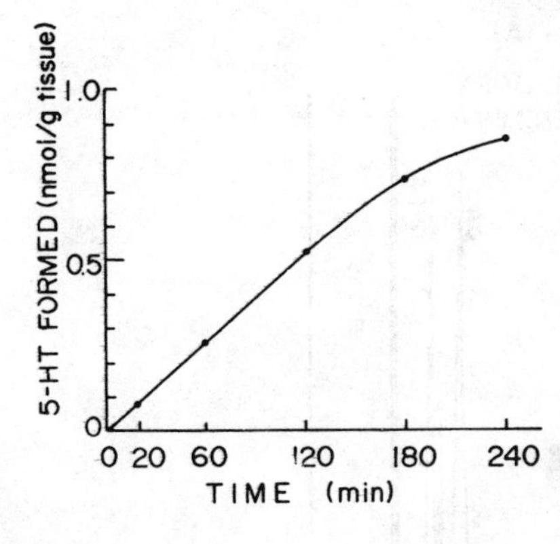

Figure 5.3 Rate of 5-HT formation using the homogenate of rat cerebral cortex as enzyme at 37°C. The incubation mixture contained 1 mg of rat cerebral cortex. (From Rahman et al., 1980.)

(*N*-M-DA) was added to each reaction mixture as an internal standard. The eluent was monitored with an electrochemical detector. The separation of these three compounds is shown in Fig. 5.2A.

The reaction mixture contained L-5-HTP as substrate, pyridoxyl phosphate as a cofactor, parglycine HCl, and the enzyme. The reaction was terminated by the addition of TCA, and after addition of the internal standard the reaction mixture was clarified by centrifugation. The sample was prepurified on Amberlite, and the 5-HT eluted and injected onto the hplc column for quantitation. The results obtained following the incubation of 5-HTP with the homogenate are shown in Fig. 5.2B. The formation of the reaction product 5-HT is indicated by the peak of this material on the chromatogram. The rate of product formation is shown in Fig. 5.3.

The enzyme was prepared from rat and human cerebral cortex. Cortical samples were homogenized in a sucrose solution, and the homogenate was used directly as the enzyme source.

5.2.3 Dopa Decarboxylase (L-Aromatic Amino Acid Decarboxylase) (Nagatsu, Oka, and Kato, 1979; D'Erme et al., 1980)

Dopa decarboxylase catalyzes the decarboxylation of L-dopa to dopamine. Pyridoxyl phosphate is a required cofactor. This enzyme has been shown to be the same protein as 5-hydroxytryptophan decarboxylase (see 5.2.2), and both are referred to by the name aromatic L-amino acid decarboxylase (AADC).

In this assay, the substrate, dopa, and the reaction product were sepa-

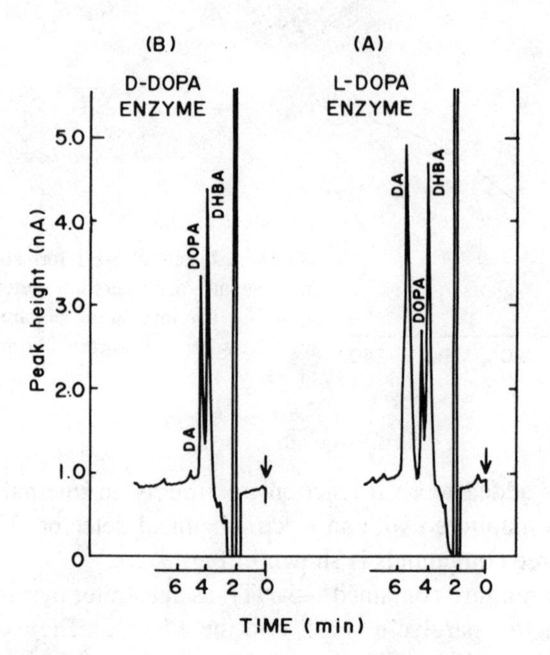

Figure 5.4 Elution pattern of AADC incubation mixtures with the homogenate of rat cerebral cortex as enzyme from hplc. The standard incubation mixture contained 0.5 mg of rat cerebral cortex. (A) Experimental incubation with L-dopa. (B) Blank incubation with D-dopa. One hundred picomoles of dihydroxybenzylamine was added to each sample after incubation. DHBA, dihydroxybenzylamine: DA, dopamine. (From Nagatsu et al., 1979b.)

rated by reverse-phase hplc and eluted isocratically with 0.1 *M* potassium phosphate buffer at pH 3.0. The eluent was monitored with an electrochemical detector. The separation obtained with this procedure is shown in Fig. 5.4B together with results obtained after incubation of L-dopa with the enzyme from rat cerebral cortex for 20 min at 37°C (Fig. 5.4A). Using a calibration curve of the type shown in Fig. 5.5, it was possible to show that 1.55 nmol of dopamine was formed. Figure 5.6 shows the rate of dopamine formation with the homogenate.

In another report the compounds were separated with a mobile phase composed of methanol–water–glacial acetic acid (25:75:1), and the eluent was monitored directly at 280 nm. The separation of dopa and dopamine using these conditions is shown in Fig. 5.7. The reaction mixture contained either dopa or α-methyldopa, an analog, as the substrate. The

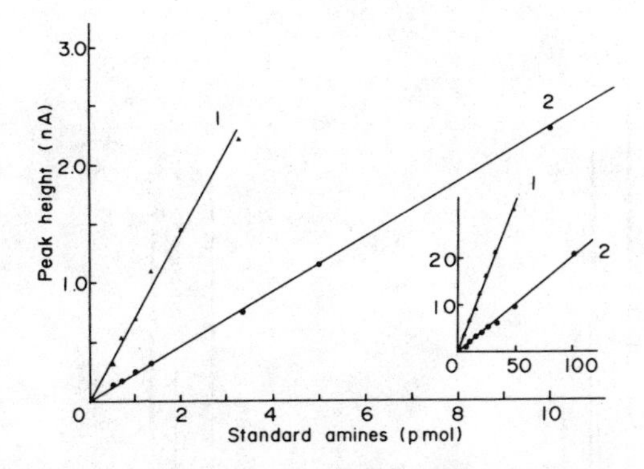

Figure 5.5 Standard curves of dopa and dihydroxybenzylamine by hplc with voltammetric detection for the peak height. One hundred microliters of a sample containing various amounts (500 fmol to 100 pmol) of dopamine (1) and hydroxybenzylamine (2) was injected into the column and was detected by a voltammetric detector. (From Nagatsu et al., 1979b.)

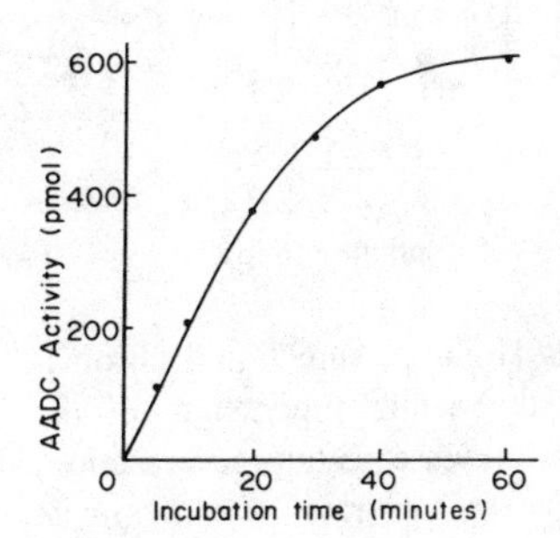

Figure 5.6 The rate of dopamine formation using the homogenate of rat cerebral cortex as enzyme at 37 °C. The incubation system contained 0.5 mg of rat cerebral cortex. (From Nagatsu et al., 1979b.)

dopamine formed after 5 min at 25 °C with the homogenate is shown in Fig. 5.8. Again, quantitation was achieved through the use of a calibration curve (see Fig. 5.9). The reaction was started by the addition of the homogenate and was terminated with 12% TCA. These solutions were clarified by centrifugation, the supernatant components were prepurified using Amberlite, and samples were injected onto the hplc column for analysis.

Samples were obtained from rat cerebral cortex. They were homogenized, and samples of the homogenate were used directly as the source of the activity.

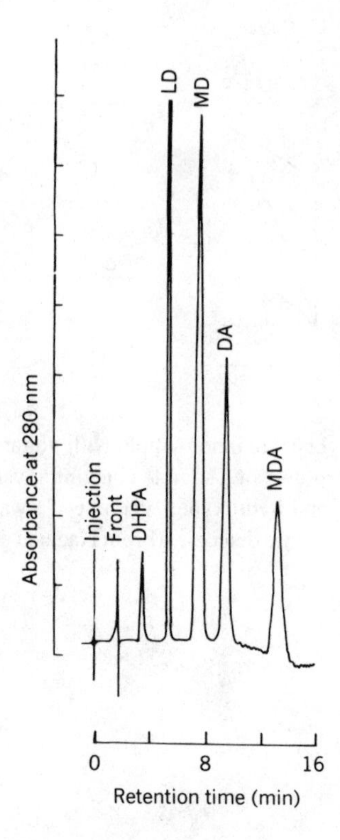

Figure 5.7 High-pressure liquid chromatogram obtained from a mixture of 3,4-dihydroxyphenylacetone (DHPA), L-dopa (LD), α-methyldopa (MD), dopamine (DA), and α-methyldopamine (MDA) under the same conditions as those used for the analysis of the enzyme-catalyzed reaction. (From d'Erme et al., 1980.)

Figure 5.8 High-pressure liquid chromatogram of the acidic supernatant of the dopa decarboxylase–catalyzed reaction, using L-dopa as substrate, after 5 min at 25 °C in the presence of 300 enzyme units. (From D'Erme et al., 1980.)

5.2.4 Dopamine β-Hydroxylase (Feilchenfeld et al., 1982)

Dopamine β-hydroxylase is the activity that catalyzes the conversion of dopamine to norepinephrine. Since these compounds are unstable, this activity is usually assayed by following the formation of octopamine from tyramine.

In one study the reactant tyramine was separated from the product octopamine by reverse-phase ion-paired hplc (μBondapak C_{18}) using a

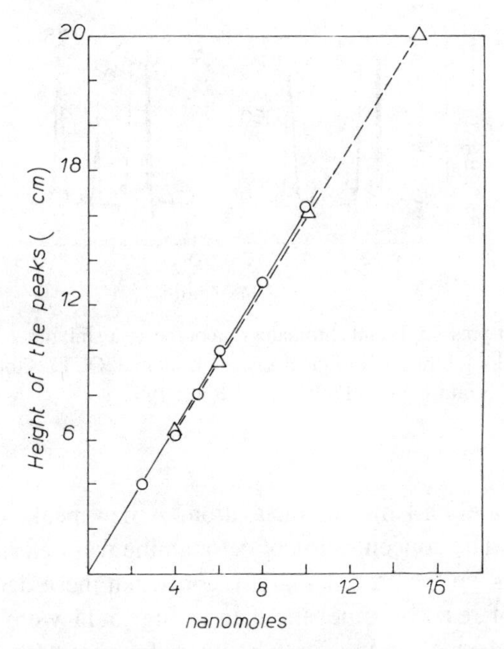

Figure 5.9 Quantitation of dopamine and dopa by hplc. The heights of the chromatographic peaks are reported as a function of the amounts of dopamine (○) or of dopa (Δ), respectively. (From D'Erme et al., 1980.)

mobile phase of 17% (v/v) methanol in H_2O containing 10 m*M* acetic acid, 10 m*M* 1-heptane-sulfonic acid (an ion-pairing reagent), and 12m*M* tetrabutylammonium phosphate. Figure 5.10A illustrates the separation of tyramine and octopamine from each other and from other components of the reaction mixture. Peaks were detected by absorbance measurements at 280 nm.

The reaction mixture contained the substrate tyramine hydrochloride (1 m*M*), sodium fumarate, ascorbic acid, catalase, and an acetate buffer at pH 5.0. The reaction was started by the addition of the activity, and samples were removed at intervals as short as 6 min and injected directly onto the hplc column for analysis. Figure 5.10A shows the chromatogram obtained before the addition of the activity. The tyramine peak is observed. (Note that the detector sensitivity has been changed midway through the elution from × 1 to × 100, an example of the "sensitivity-shift" technique described in Chapter Three.) Figure 5.10B shows a chromatogram

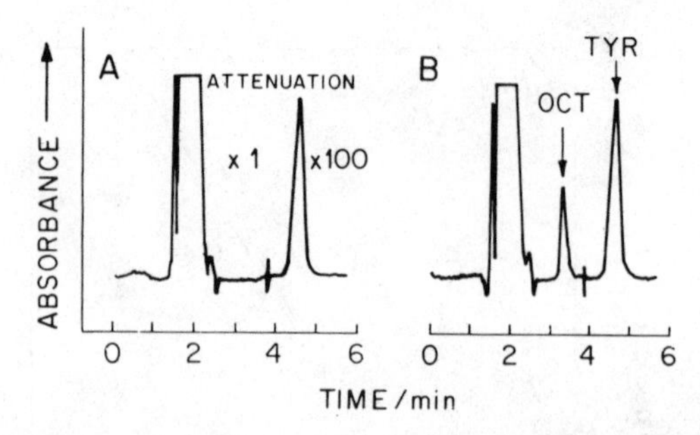

Figure 5.10 High-pressure liquid chromatogram of the assay mixture (A) prior to and (B) 13.65 min after addition of DBH to the assay stock sample. OCT, octopamine (3.2 min); TYR, tyramine (4.5 min). (From Feilchenfeld et al., 1982.)

after approximately 14 min of incubation. A new peak, octopamine, is clearly visible. The concentration of octopamine was determined from the area of the peak on the chromatogram, and when these data were plotted as a function of reaction time, the data in Fig. 5.11 were obtained. The rate of product formation is seen to be linear for about 12 min and 9 min, respectively, for two concentrations of enzyme obtained from a commercial source.

5.2.5 Catechol *O*-Methyltransferase (Pennings and Van Kempen, 1979; Koh et al., 1981)

The enzyme *O*-methyltransferase catalyzes the methylation of catecholamines and their metabolites. Two assay methods have been developed. In both, *S*-adenosyl-L-methionine was the methyl donor. In one study norepinephrine was the substrate, while products of the reaction, nor*metanephrine* and nor*para*nephrine, were converted to the more stable and more easily obtained compounds vanillin and isovanillin, respectively, by periodate oxidation. The oxidation also allowed for the extraction of the incubation mixture with organic solvents such as ethyl acetate allowing for a more complete deproteinization.

In this study (not shown), the compounds vanillin and isovanillin together with *p*-hydroxyacetanilide, added as an internal standard, were separated by reverse-phase hplc (LiChrosorb) with a methanol–50 m*M* phosphate buffer (pH 7.2) (3:7, v/v) as the mobile phase. The compounds

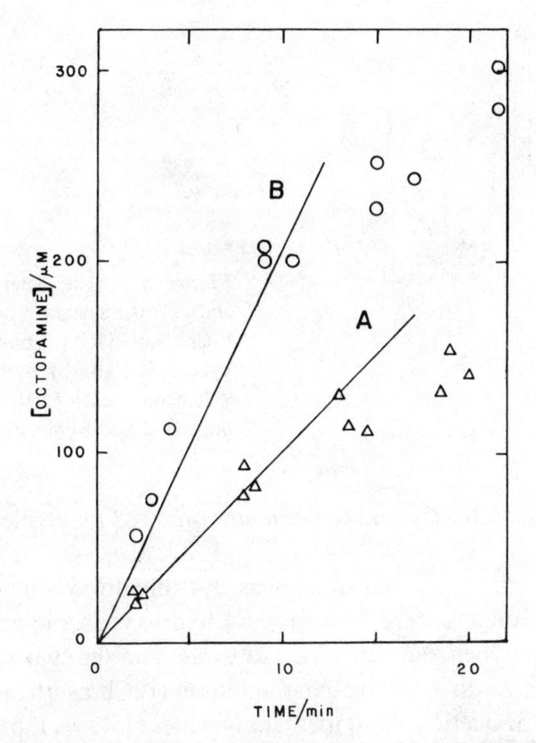

Figure 5.11 Time course of production of octopamine with varying DBH concentrations. (A, Δ) [DBH] = 0.01 mg/mL; (B, ○) [DBH] = 0.02 mg/mL. Lines represent the least-squares fits of the initial data. (From Feilchenfeld et al., 1982.)

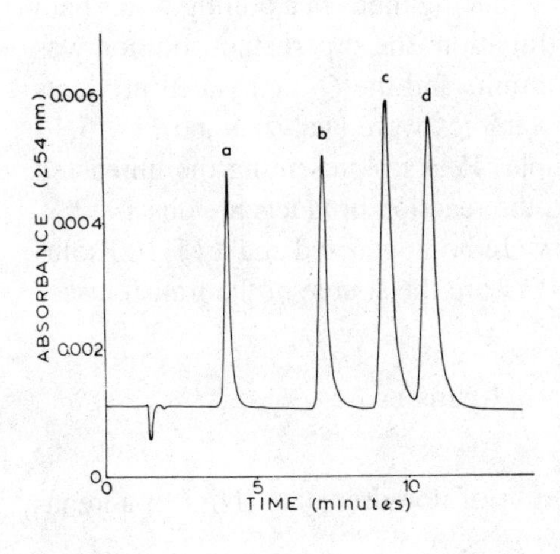

Figure 5.12 Optimal separation of (a) DHBA, (b) 4-hydroxybenzoic acid (c) MHBA, and (d) HMBA, respectively, on a LiChrosorb 5 RP 18 column. (From Pennings and Van Kempen, 1979.)

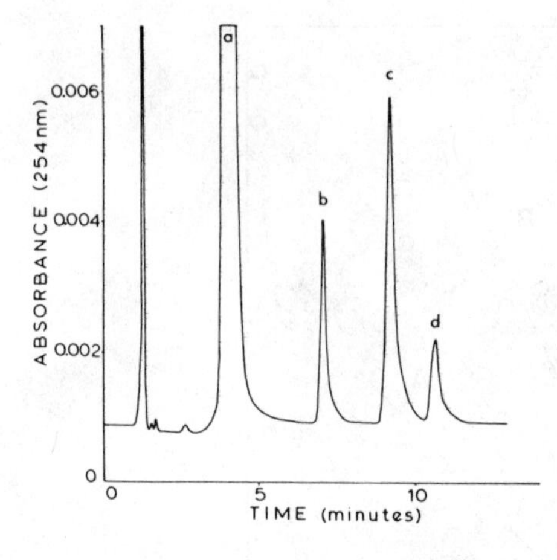

Figure 5.13 The determination of *m*- and *p*-*O*-methylated products (c,d) on a LiChrosorb 5 RP 18 column. (a) 4,hydroxybenzoic acid, (b) 3-methoxy-4- hydroxybenzoic acid (MHBA). (From Pennings and Van Kempen, 1979.)

were eluted isocratically and the eluent monitored by an electrochemical detector.

In another study, the substrate was 3,4-dihydroxybenzoic acid, and the reaction products were 3-methoxy-4-hydroxybenzoic acid and 3-hydroxy-4-methoxybenzoic acid. The substrate and the two products were separated by hplc on a reverse-phase column (LiChrosorb) with a mobile phase of 0.05 *M* acetic acid in methanol–water (1:4, v/v), pH 3.2. Figure 5.12 shows the separation obtained under these conditions.

The reaction mixture was contained in a volume of 1 mL Tris-HCl buffer (pH 7.9), *S*-adenosylmethionine, $MgCl_2$, 3,4-dihydroxybenzoic acid, and dithiothreitol. Reactions were started by the addition of the activity and terminated after 10 min by placing tubes in a boiling water bath for 2 min. After cooling and centrifugation, the supernatant solution was loaded onto a DEAE-Sephadex column, and the *O*-methylated products were eluted with 0.075 m*M* HCl. Samples were analyzed and Fig. 5.13 shows the chromatogram of a sample. Peaks representing the unreacted substrate, an internal standard, and the reaction products are observed.

Samples obtained from rat liver were homogenized and an S-100 solution prepared. Samples of this S-100 were the source of the transferase.

5.2.6 Phenylethanolamine N-Methyltransferase (Trocewicz et al., 1982)

This enzyme catalyzes the conversion of noradrenaline (NA) to adrenaline (AD).

The assay, which measured only the amount of AD formed, used ion-paired reverse-phase hplc chromatography. The separation was carried out on a C-18 (Nucleosil) column with a mobile phase of 0.1 *M* sodium phosphate buffer (pH 2.3–3.5) containing 5 m*M* sodium pentanesulfonate as the counterion to form ion pairs with the catecholamines, and 0.5% (v/v) acetonitrile. The separation of NA from AD is shown in Fig. 5.14.

The reaction mixture (250 μL) contained parglycine to inhibit secondary reactions catalyzed by monoamine oxidase activity, *S*-adenosylmethionine as the donor, and NA as the acceptor. Dihydroxybenzylamine

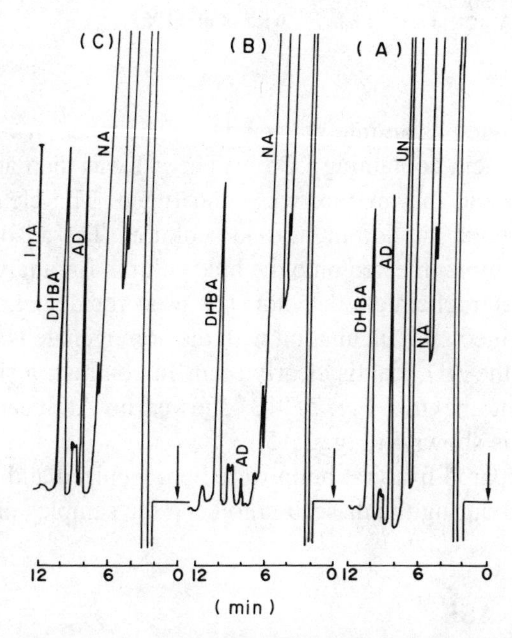

Figure 5.14 A typical elution pattern of phenylethanolamine *N*-methyltransferase incubation mixtures with the homogenate of rat pons plus medulla oblongata as enzyme. The incubation mixture contained 10 mg of rat pons plus medulla oblongata as enzyme and 16 μ*M* noradrenaline (NA) and 18 μ*M* *S*-adenosyl-L-methionine (SAM) as substrates. (A) Experimental incubation with homogenate of 10 mg of rat pons plus medulla oblongata. (B) Blank incubation without enzyme. (C) 15 pmol of adrenaline (AD) were added to another blank incubation after stopping the reaction. Formation of 16.6 pmol of AD from NA during 60 min incubation at 37°C was calculated from a calibration curve. DHBA = dihydroxybenzylamine (internal standard), UN = unknown peak. (From Trocewicz et al., 1982.)

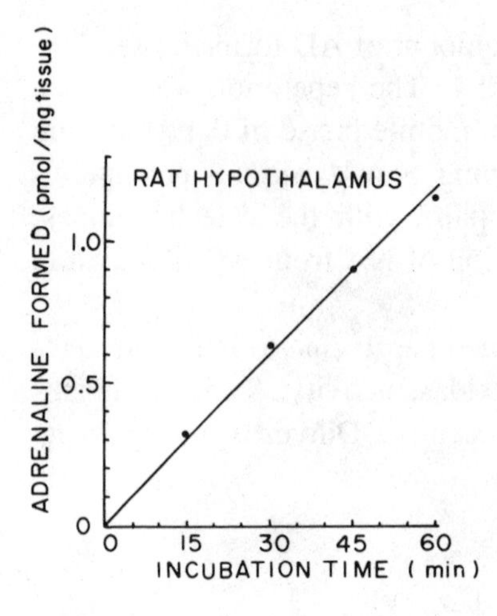

Figure 5.15 The rate of adrenaline formation using an homogenate of rat hypothalamus as enzyme at 37°C. Standard incubation system containing 10 mg of tissue was used. (From Trocewicz et al., 1982.)

(DHBA) was added as an internal standard. The reaction was terminated with perchloric acid containing EDTA. The pH was then adjusted to 8.5 with Tris-HCl, and the mixture was centrifuged. The clear supernatant was passed through an aluminum oxide column. The adsorbed AD was eluted, and a sample injected onto the hplc column for analysis and quantitation. An electrochemical detector was used for detection. The chromatogram obtained after incubation with the homogenate is shown in Fig. 5.14A, where the AD peak is clearly seen. In contrast, a similar incubation but without enzyme (Fig. 5.14B) showed no AD peak. The rate of AD formation is shown in Fig. 5.15.

Samples prepared from rat brain were homogenized and used directly. Care was taken during the dissection to keep the samples on ice.

5.3 PROTEINASE

5.3.1 Vertebrate Collagenase (Gray and Saneii, 1982)

In both the $alpha_1$ and $alpha_2$ chains of Type I collagen, degradation of the protein begins with cleavage of the Gly-Ile or the Gly-Leu bond by collagenase.

In the assay developed for this activity, the collagen was replaced by the peptide DNP-Pro-Gln-Gly-Ile-Ala-Gly-Gln-D-Arg. During the course of the reaction this substrate was cleaved into the two products DNP-Pro-

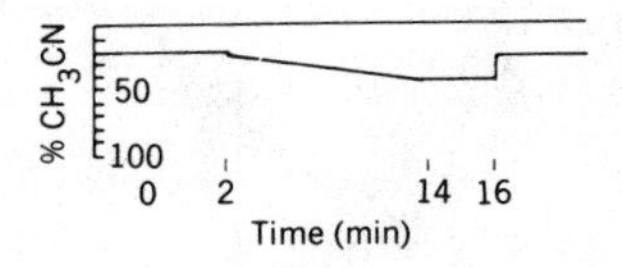

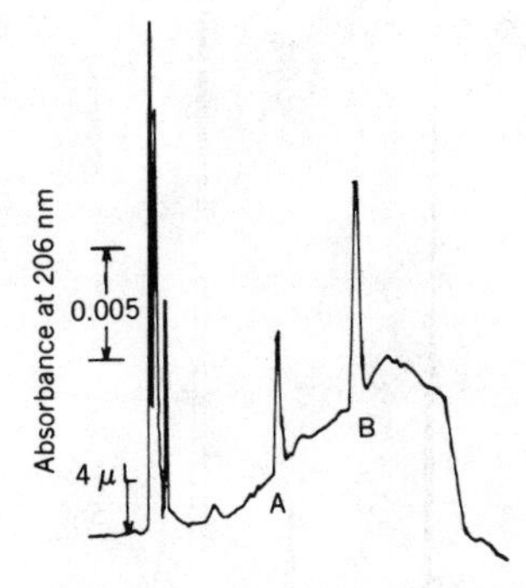

Figure 5.16 Separation of DNP-Pro-Gln-Gly (A) from DNP-Pro-Gln-Gly-Ile-Ala-Gly-Gln-D-Arg (B) using reverse-phase hplc. An aliquot of 4.0 μL was injected at the arrow. The initial solvent was 0.1% H_3PO_4/CH_3CN (80:20). The column was eluted with this solvent for 2.0 min, following which the CH_3CN was increased to 40% linearly over a 12 min time period. The flow rate was 2.0 mL/min. The upper graph shows the gradient used. (From Gray and Saneii, 1982.)

Gln-Gly and Ile-Ala-Gly-Gln-D-Arg. The substrate and the two products were separated on a reverse-phase column (Varian MCH-10) eluted as follows: Initially the column was equilibrated with a mobile phase consisting of 0.1% H_3PO_4–CH_3CN (80:20, v:v), which was followed by a mobile phase in which the proportion of CH_3CN was increased linearly up to 40% by volume. This gradient resulted in the elution of both the DNP–tripeptide reaction product and the unreacted substrate, as shown in Fig. 5.16. The eluent was monitored at 206 nm, although monitoring for the DNP derivative at 365 nm was also feasible.

The reaction mixture included Tris-HCl buffer, NaCl, $CaCl_2$, and the DNP-octapeptide substrate. The reaction, initiated by the addition of enzyme, was incubated at 37°C. The total volume of the reaction mixture varied between 10 and 25 μL depending on the number of samples to be analyzed. Successive aliquots of 3–10 μL were removed with a syringe at timed intervals and applied directly to the column without further treatment. [The column was protected by a guard column packed with pellicular packing (V/dac Reverse Phase Pellicular Packing).] The results of incubation of the substrate with tadpole collagenase are shown in Fig. 5.17. Chromatograms for three time points are shown. In Fig. 5.18, the rate of hydrolysis is shown as a function of enzyme concentration. Because the gradient had to be reversed prior to injection of a new sample, the interval between samples was 18–20 min.

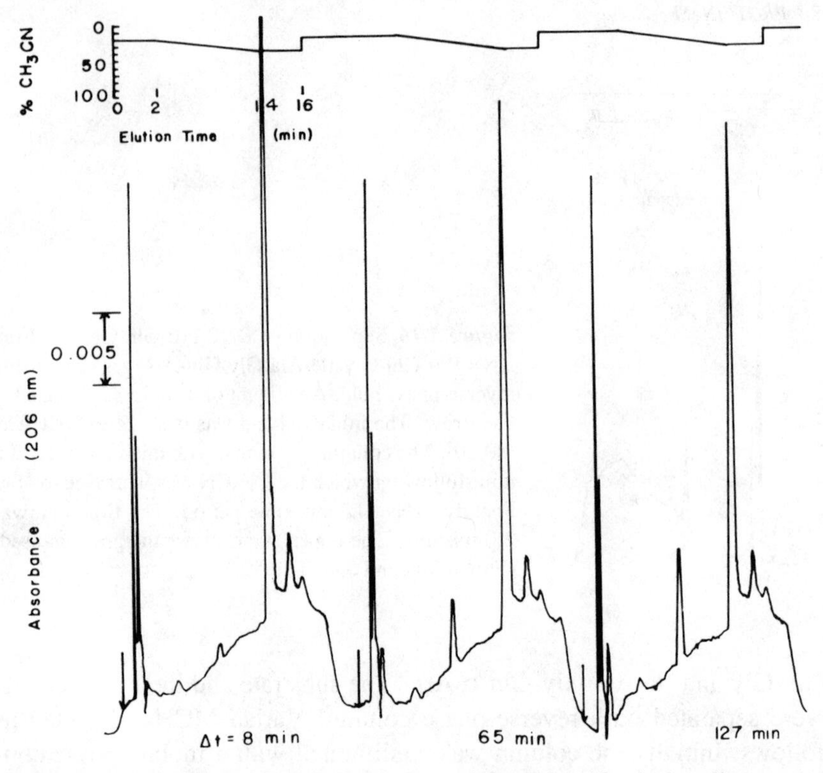

Figure 5.17 Separation of hydrolysis products of the action of tadpole collagenase on DNP-Pro-Gln-Gly-Ile-Ala-Gly-Gln-D-Arg by reverse-phase hplc. The upper tracing shows the gradient employed. The reaction was initiated at zero time by the addition of 4 μg of enzyme. At the indicated times, a 4.0 μL aliquot of the reaction mixture was injected onto the hplc column and the peptides separated. (From Gray and Saneii, 1982.)

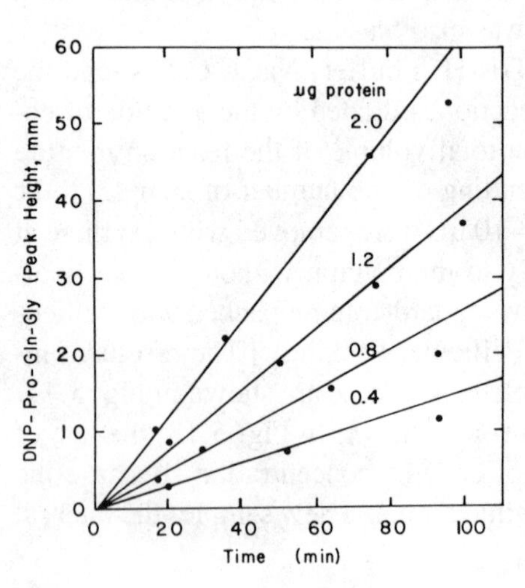

Figure 5.18 Dependence of rate of hydrolysis of DNP-Pro-Gln-Gly-Ile-Ala-Gly-Gln-D-Arg on collagenase concentration. Substrate concentration was 1.8 m*M*, pH 7.6, 37°C. (From Gray and Saneii, 1982.)

The enzyme vertebrate collagenase was partially purified from a lyophilized tissue culture medium of back skin from tadpoles. The medium was harvested, and the protein precipitated by 40% ammonium sulfate collected by centrifugation. The precipitate was recovered and purified further on G-200 Sephadex.

5.3.2 Dipeptidyl Carboxypeptidase (Angiotensin I Converting Enzyme, EC 3.4.15.) (Baranowski et al., 1982)

Angiotensin I converting enzyme converts angiotensin I to angiotensin II and the dipeptide histidyl-leucine (His-Leu).

The reactant was separated from the product as the fluorescamine derivatives on a reverse-phase hplc column (Partisil ODS) and eluted isocratically using a two-solvent system. To elute the dipeptide the solvent was 60% acetonitrile in H_2O diluted 9:1 with 1 *M* acetic acid, at a final pH of 3.5 (Fig. 5.19). To elute the angiotensin compounds, 38%

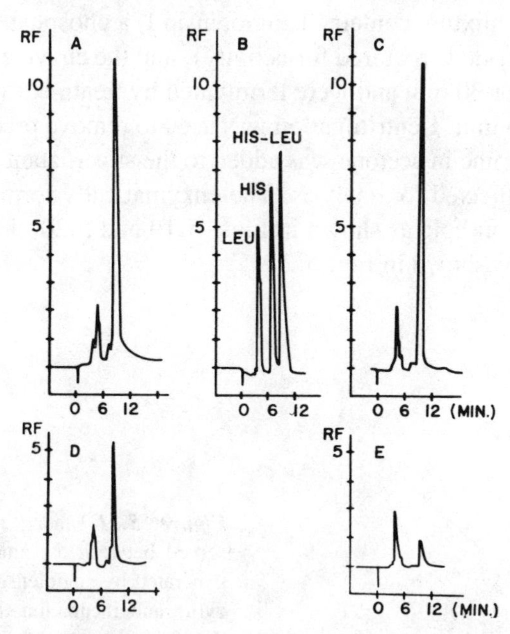

Figure 5.19 Hplc separation of a standard mixture of fluorescamine derivatives of histidine, leucine, and histidyl-leucine (B). Hplc separation of enzymatically formed peptide fragments in the presence (A, C) and absence (D,E) of sodium chloride in the incubation mixture. (From Baranowski et al., 1982.)

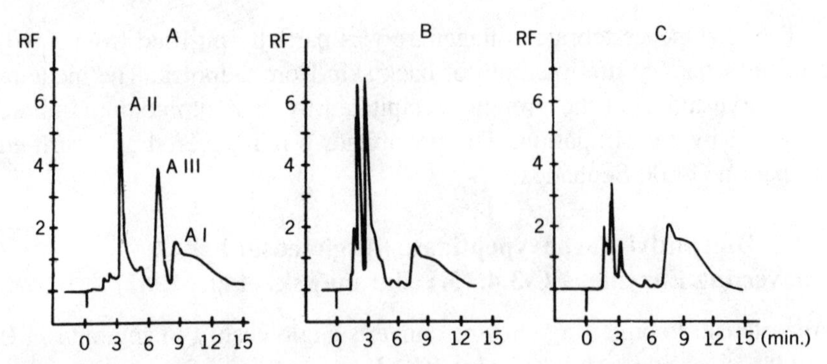

Figure 5.20 Hplc separation of a standard mixture of the fluorescamine derivatives of angiotensins I, II, and III (A). Hplc separation of enzymatically formed peptide fragments in the presence (B) and absence (C) of sodium chloride in the incubation mixture. (From Baranowski et al., 1982.)

acetonitrile in 1 *M* ammonium acetate (pH 4.0) was used (Fig. 5.20). The eluent was monitored with a fluorometer.

The reaction mixture contained angiotensin I, a phosphate buffer at pH 8.0, NaCl (chloride is required for activity), and the enzyme. Incubations were at 37°C for 30 min and were terminated by treatment with a boiling water bath for 5 min. Centrifugation was used to remove precipitated protein. Fluorescamine in acetone was added to the supernatant solution, and samples were injected for analysis. The enzymatically formed fragments were separated on hplc as shown in Figs. 5.19 and 5.20; the rate of His-Leu formation is shown in Fig. 5.21.

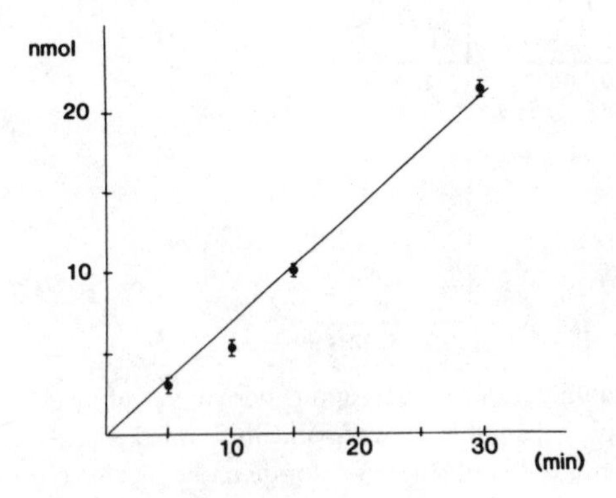

Figure 5.21 Linear relationship observed between the amount of His-Leu generated by angiotensin-converting enzyme and incubation time. Number of determinations are: 3, 5, 5, and 5 for 5, 10, 15, and 30 min, respectively. (From Baranowski et al., 1982.)

The dipeptidyl carboxypeptidase was prepared from rat lungs. A microsomal fraction was prepared and extracted with detergent (sodium deoxycholate) and clarified by centrifugation. The supernatant solution was dialyzed against sodium phosphate, and the dialysate was stored frozen.

5.3.3 Luteinizing Hormone–Releasing Hormone Peptidase (Advis et al., 1982)

The luteinizing hormone-releasing hormone peptidase is one of many peptidases that catalyze the hydrolysis of neuropeptides.

The assay used for this activity was based on the separation of the substrate, luteinizing hormone-releasing hormone (LHRH) from its degradation product, a pentapeptide. A reverse-phase hplc (C-18) column was used, and LHRH was eluted isocratically in an aqueous mobile phase composed of 42% acetonitrile containing 0.7 m*M* tetraethylammonium phosphate. To analyze for degradation products, a mobile phase of 0.1 *M* sodium phosphate–0.1 *M* H_3PO_4 (pH 2.5) was used to equilibrate the C-18 column. The reaction fragments were eluted with an exponential gradient of 0–30% acetonitrile. Several fragments generated by the reaction include $LHRH_{1-5}$, $LHRH_{6-10}$, and $LHRH_{1-3}$, and all were separated and collected.

The LHRH peptidase activity was assayed with LHRH in a phosphate buffer (pH 7.2). The reaction was started by addition of LHRH and terminated by heating at 110°C for 6 min in a heating apparatus (Reacti-Thermi, Pierce). The sample was clarified by centrifugation and stored prior to activity determinations. The chromatograms obtained at zero time and after a 2-h incubation are shown in Fig. 5.22. The hypothalamus from the female rat was sampled using a "punch" technique. Samples were homogenized in ice-cold phosphate buffer at pH 7.4 and the lysate was centrifuged for 5 min at 10,000 *g* at room temperature. The supernatant fraction was used as the source of peptidase activity and stored at − 80°C.

5.3.4 Papain Esterase (Chen et al., 1982)

Papain is a proteolytic enzyme from plants. In the hplc assay developed to measure its activity, the water-soluble *N*-benzyl-L-arginine ethyl ester (BAEE) is used as the substrate.

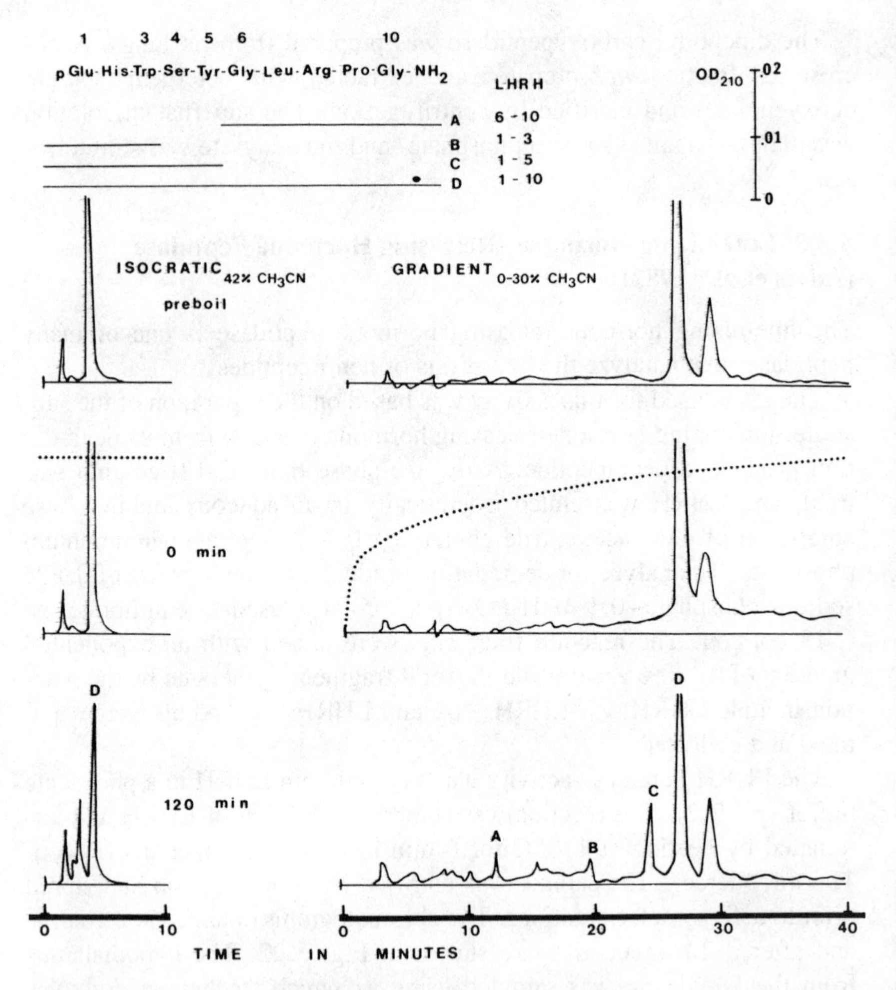

Figure 5.22 Isocratic and gradient hplc analysis of median eminence (ME) supernatant fraction after incubation with LHRH. Total peptidase activity was assessed in ME and was analyzed by hplc. The upper hplc tracing represents a preboiled control, the middle tracing a zero time incubation, and the lower a 2 h preincubation. The peptide fragments are indicated in the upper panel by letter (A–D) and the corresponding peaks they represent are shown by the same letter. (From Advis et al., 1982.)

The separation of the substrate from the reaction product benzoyl-arginine was carried out by reverse-phase hplc on a μ-Bondapak CN column with a mobile phase of 0.05 *M* ammonium acetate–methanol (85:15). The column was eluted isocratically, and the eluent was monitored at 254 nm.

The reaction mixture contained the substrate BAEE, mercaptoethanol, EDTA, and sodium chloride. The reaction was started by the addition of papain, and the immediate adjustment of the incubation mixture to pH 6.2. After 5 min, the reaction was terminated by the addition of 30% acetic acid. Precipitated material was removed by filtration, using a 0.45 μm Millipore filter, and the filtrate was analyzed by hplc. The results of an assay are shown in Figure 5.23, where the appearance of the reaction product benzoyl-arginine is evidence of enzymatic activity.

The papain was obtained from commercial sources.

NOTES

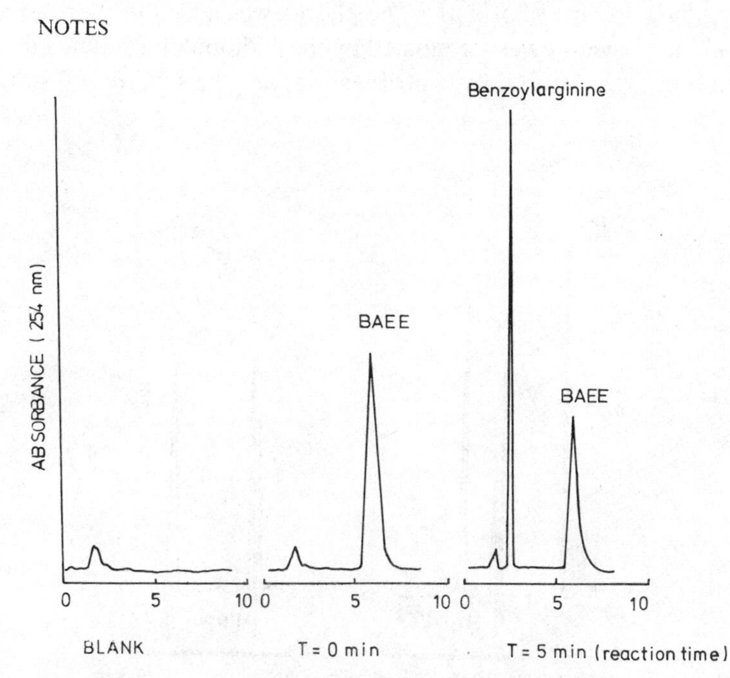

Figure 5.23 Hplc profile of papain-catalyzed BAEE hydrolysis product. Conditions: column, μBondapak CN; eluent, 0.05 *M* ammonium acetate–methanol (85:15); flow rate, 2.0 mL/min; detector, 254 nm at ambient temperature. (From Chen et al., 1982.)

5.3.5 Plasma Carboxypeptidase N (Kininase I, Bradykinin Destroying Enzyme, EC 3.4.12.7) (Marceau et al., 1983)

Plasma carboxypeptidase N degrades and therefore inactivates bradykinin. This activity may have a role in the regulation of inflammatory peptides. The hplc method developed for its assay uses the dipeptide hippuryllysine (Hip-Lys) as the substrate and measures activity by measuring the release of hippuric acid.

The separation of substrate from the product was carried out by reverse-phase hplc (C-18 μBondapak) using a mobile phase of a 1:4 mixture of methanol and 0.001 *M* K_2HPO_4 and H_3PO_4 (pH 3.0). The column was protected by a precolumn packed with Corasil. The column was eluted isocratically, and detection was at 230 nm. The separations obtained under these conditions are shown in Fig. 5.24.

The reaction mixture contained HEPES (N-2-hydroxyethylpiperazine-N^1-2-ethanesulfonic acid) buffer (pH 7.75) with NaCl and the substrate Hip-Lys in a volume of 500 μL. The plasma was added to start the reaction, and the reaction was terminated by the addition of absolute ethanol. The mixture was clarified by centrifugation, and the supernatant solution

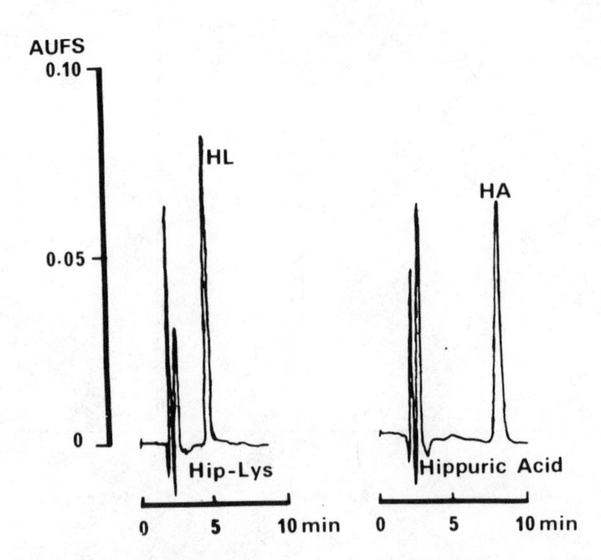

Figure 5.24 Hplc traces of standard solutions (32 μg/mL) of Hip-Lys (HL) and hippuric acid (HA). (From Marceau et al., 1983.)

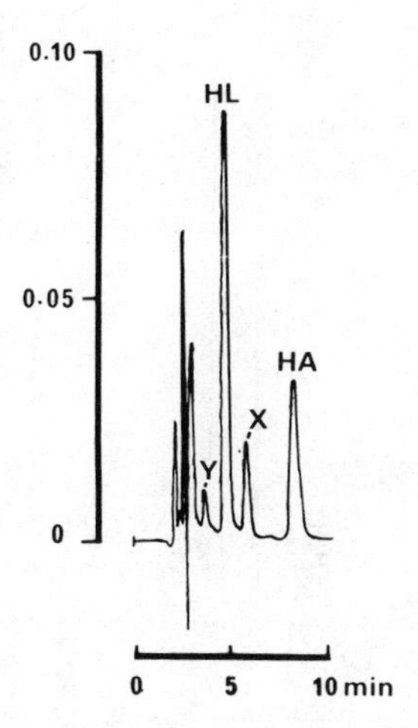

Figure 5.25 Hplc traces of solutions of Hip-Lys containing 75% human plasma incubated for 15 min. The substrate concentration was 1.02 m*M*. X and Y represent unknown substances from plasma. (From Marceau et al., 1983.)

applied to the hplc for analysis. The result of a 15-min incubation is shown in Fig. 5.25.

The carboxypeptidase was prepared from plasma.

5.3.6 Dipeptidase (Horiuchi et al., 1982)

In this study, the tripeptide hippurylhistidylleucine (Hip-His-Leu) was used as a substrate, and the assay involved measuring the amount of hippuric acid released by the enzyme.

The substrate was separated from the product by reverse-phase hplc (Nucleosil 7 C-18) using a mobile phase of methanol–10 m*M* KH_2PO_4 (1:1) adjusted to pH 3.0 with phosphoric acid. Detection was at 228 nm. The separation of the components of the reaction mixture is shown in Fig. 5.26.

The reaction mixture contained a phosphate buffer at pH 8.3 with NaCl and the substrate Hip-His-Leu. The reaction was incubated with the dipeptidase preparation for 30 min at 37°C and terminated with 3% metaphosphoric acid. The mixture was centrifuged, and a sample of the supernatant solution was injected onto the hplc column for analysis. The results

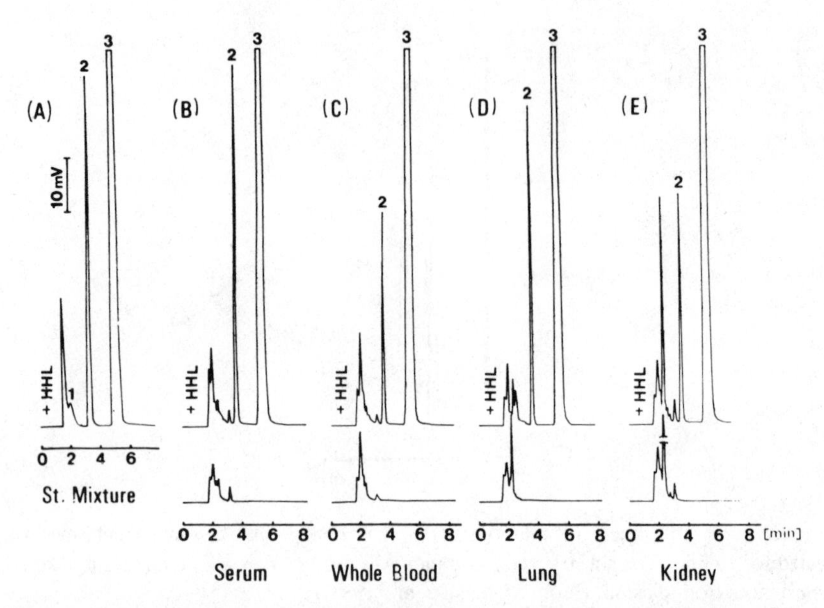

Figure 5.26 Chromatograms obtained from various samples incubated with (upper diagram) or without (lower diagram) Hip-His-Leu (HHL). (A) Standard mixture of 2.7 nmol His-Leu, 2.7 nmol hippuric acid, and 100 nmol Hip-His-Leu. (B) A 50 μL aliquot of serum or (C) whole blood was incubated with or without 5 m*M* Hip-His-Leu. After 30 min, 0.75 mL of 3% *meta*phosphoric acid was added and centrifuged. (D) Lung or (E) kidney was homogenized in 5 volumes of chilled Tris-HC1 buffer containing 0.5% Nonidet-P40, and centrifuged. The supernatant was incubated with or without 5 m*M* Hip-His-Leu. In the case of lung, the supernatant was diluted 20 times with the buffer prior to incubation with Hip-His-Leu. Peaks: (1) His-Leu, (2) hippuric acid, (3) Hip-His-Leu. (From Horiuchi et al., 1982.)

of an assay are shown in Fig. 5.26. The appearance of the hippuric acid is taken as evidence of enzymatic activity. The rate of product formation is shown in Fig. 5.27.

Peptidase preparations were obtained from rat blood, lung, and kidney. The latter were chopped and homogenized and clarified by centrifugation at 20,000g for 20 min. The supernatant solution was used as the source of peptidase activity.

5.3.7 Aminopeptidase (Mousa and Couri, 1983)

Aminopeptidases are involved in the metabolism of opioid peptides including enkephalins and β-endorphins. An hplc method has been devel-

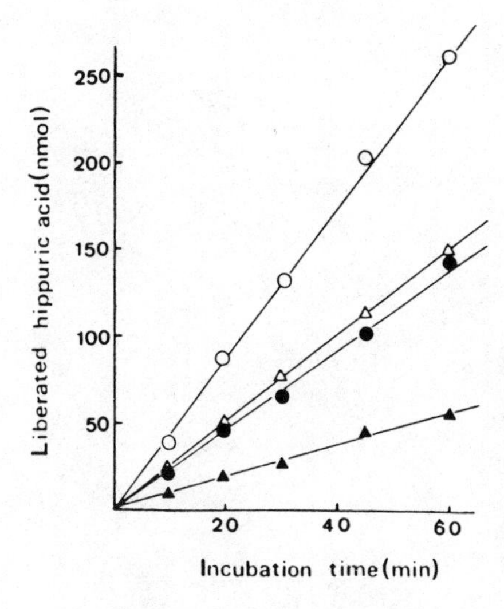

Figure 5.27 Dependence on incubation time of hippuric acid formation from Hip-His-Leu catalyzed by serum (○), whole blood (●), lung (Δ), and kidney (▲). A 50 μL aliquot of sample was used. Each point represents the mean of three determinations. (From Horiuchi et al., 1982.)

oped to measure the hydrolysis of these compounds by measuring the formation of tyrosylglycylglycine, tyrosylglycine, and tyrosine.

The separation was accomplished by reverse-phase hplc (Ultrasphere ODS column) with a mobile phase of 50 m*M* sodium phosphate buffer (pH 2.1) with acetonitrile and methanol (90:5:5). The column was eluted isocratically, and an electrochemical detector was used. The separation of the standards is shown in Fig. 5.28.

The reaction mixture for peptidase activity contained the substrate methionine enkephalin with Tris-HCl (pH 7.4) as buffer. The reaction was started by the addition of serum peptidase, and after various intervals the reaction was terminated by adding 1 *M* HCl.

The aminopeptidase was obtained from either serum, rat brain or synaptosomal membrane.

5.4 AMINO ACID METABOLISM

5.4.1 Ornithine Aminotransferase (O'Donnell et al., 1978)

Ornithine aminotransferase (OATase) is a mitochondrial matrix enzyme that catalyzes the primary reaction

$$\text{L-ornithine} + \alpha\text{-ketoglutarate} \rightarrow$$
$$\text{glutamic-}\gamma\text{-semialdehyde} + \text{glutamate.}$$

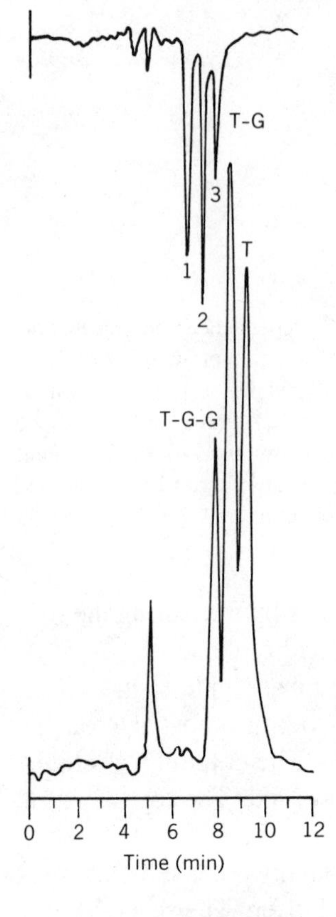

Figure 5.28 Representative chromatogram of tyrosine (T), tyrosylglycine (T-G) and tyrosylglycylglycine (T-G-G) standards. The upper chromatogram (hplc-UV) and the lower chromatogram [hplc-electrochemical detection (ED)] show the separation of 55 ng of T-G-G (1), 70 ng of T-G (2), and 35 ng of T (3). Analytes were detected at 205 nm (0.1 A) and + 1.25 V oxidation potential (100 nA). (From Mousa and Couri, 1983.)

However, following its formation, the semialdehyde undergoes a spontaneous cyclization and is converted to a Δ^1-pyrroline-5-carboxylic acid (P5C), a proline precursor.

The assay developed for this activity involves the reaction of the P5C with *O*-aminobenzaldehyde (OAB) to form the reaction product dihydroquinozolinium (DHQ). The DHQ and unreacted OAB were separated by reverse-phase hplc (LiChrosorb C18). The column was eluted isocratically with a mobile phase of 1 part methanol to 2 parts water, and the separation shown in Fig. 5.29 obtained. The reaction mixture contained L-ornithine (35 m*M*), α-ketoglutarate, potassium phosphate (pH 7.4), and pyridoxyl phosphate in a total volume of 2 mL. The reaction was started by the addition of the homogenate and terminated by the addition of 1 mL

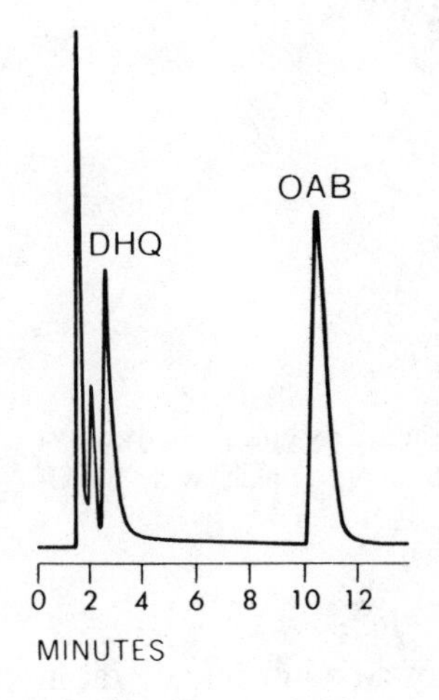

Figure 5.29 Isocratic reverse-phase hplc/chromatogram showing separation and detection of dihydroquinozolinium (DHQ) and *o*-aminobenzaldehyde (OAB). The column was LiChrosorb C-18, and 4.6 times 250 mm, and the solvent system was methanol–water (1:2), pumped at 1.5 mL/mm. Detection was at 254 nm with an Altex 310 chromatograph. (From O'Donnell et al., 1978.)

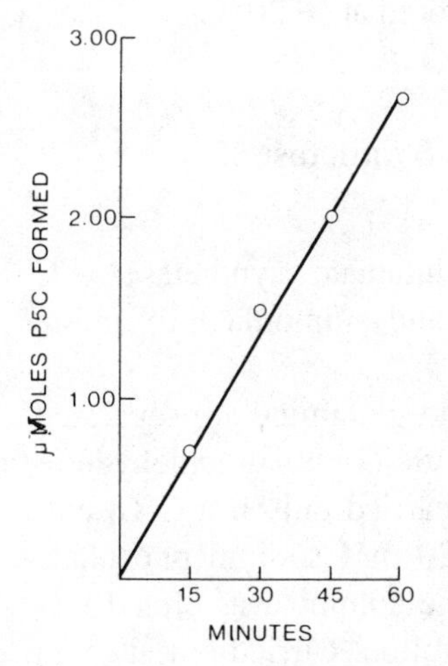

Figure 5.30 OTA activity as determined by hplc plotted against time. The reaction was linear for 60 min. (From O'Donnell et al., 1978.)

of 3 *N* HCl containing the OAB. Precipitated protein was removed by centrifugation (3000 rpm), and samples of the supernatant solution (10 μL) were injected for analysis.

Quantitation was by means of peak height rather than peak area. The OTA activity was determined over a 60-min incubation, and, as shown in Fig. 5.30, it was linear during this interval as well as proportional to the amount of homogenate added (Fig. 5.31).

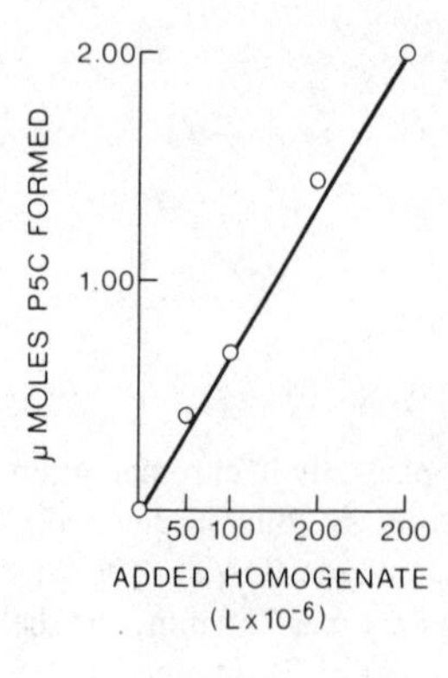

Figure 5.31 OTA activity determined by hplc plotted against added homogenate. The reaction was linear with homogenate. (From O'Donnell et al., 1978.)

The homogenate was prepared from the liver of an adult female rat in 20% sucrose containing potassium phosphate (pH 7.4). The homogenate was centrifuged at 1000 rpm for 15 min and stored at $-20°C$.

5.4.2 Glutamine Synthetase and Glutamate Synthetase (Martin et al., 1982)

Glutamine synthetase (EC 6.3.1.2) and glutamate synthetase (EC 1.4.7.1) catalyze the assimilation of nitrate and ammonia into amino acids.

In this hplc assay method, glutamic acid and glutamine are separated by reverse-phase hplc after derivatization with *O*-phthaldialdehyde/2-mercaptoethanol (OPA). The separation was carried out on a C-18 column (μBondapak) using a mobile phase of 20 m*M* sodium phosphate buffer (pH 6.8) and methanol (64:36 v:v). The column was eluted isocratically and monitored at 340 nm. The separation, carried out at room temperature, is shown in Fig. 5.32. The reaction mixtures were prepared according to previously published procedures. The reactions were terminated using either chloride or TCA. Precipitates were removed by centrifugation, and the supernatant solution was used for the assay. Figure 5.33 shows chromatographic profiles obtained after incubation for 15 min. The formation of glutamine and glutamate are shown.

The enzymes were prepared either from root nodules or from rice leaves by previously published procedures.

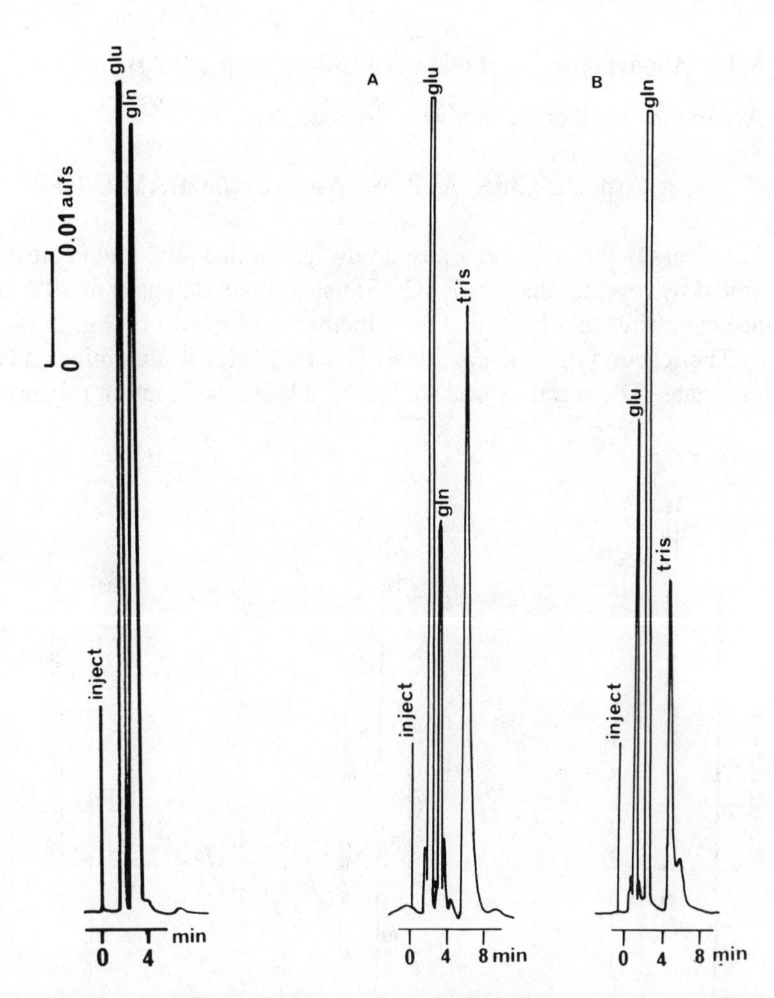

Figure 5.32 Chromatograms of *o*-phthaldialdehyde, glutamic acid, and glutamine standards. Sample contained 5 nmol of glutamic acid and 10 nmol of glutamine. (From Martin et al., 1982.)

Figure 5.33 Chromatograms of enzyme assay media. (A) Elution profile of the assay medium of glutamine synthetase. Alder root nodule enzyme plus assay mixture was incubated for 15 min. (B) Elution profile of the assay medium of glutamate synthetase. Rice leaves enzyme plus assay mixture was incubated for 15 min. (From Martin et al., 1982.)

5.4.3 Asparagine Synthetase (Unnithan et al., 1984)

Asparagine synthetase catalyzes the reaction

$$\text{L-Asp} + \text{L-Gln} + \text{ATP} \rightarrow \text{L-Asn} + \text{L-Glu} + \text{AMP} + \text{PP}_i$$

In this assay the asparagine, aspartate, glutamine, and glutamate are separated by reverse-phase hplc (C-18) using a mobile phase of 70% sodium acetate buffer (pH 5.9), and 30% methanol as shown in Fig. 5.34.

The reaction mixture contained Tris-HCl, MgCl, glutamine, ATP, and aspartate. The reaction was started by adding the reaction mixture cock-

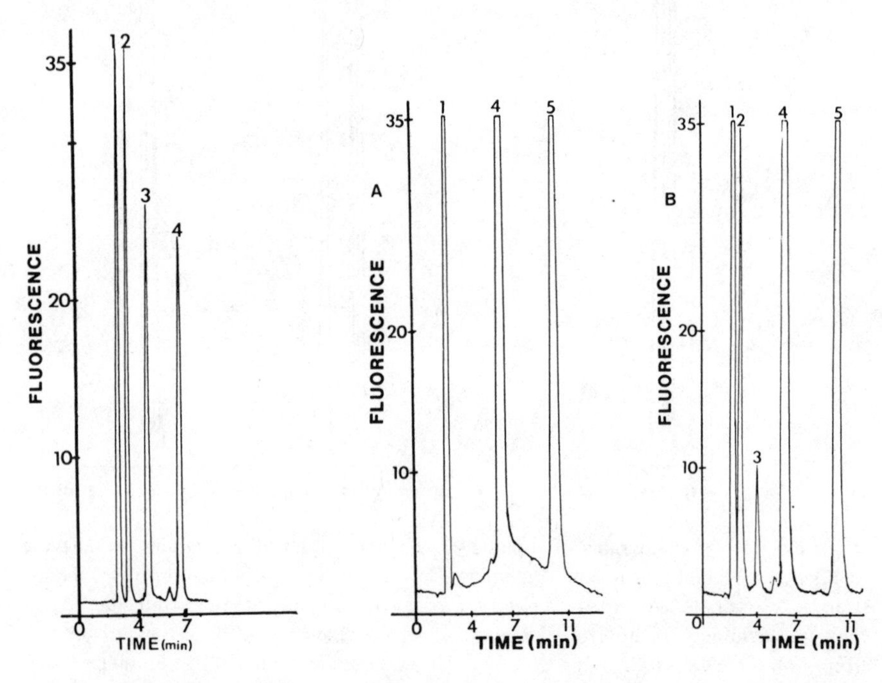

Figure 5.34 Chromatogram of (1) *o*-phthaldialdehyde aspartate, (2) glutamate, (3) asparagine, and (4) glutamine. A mixture of 0.025 m*M* of each amino acid was made and 20 μL injected. (From Unnithan et al., 1984.)

Figure 5.35 Chromatography of enzyme assay media. The peaks are (1) aspartate, (2) glutamate, (3) asparagine, (4) glutamine, (5) Tris-HCl buffer. Elution profile of the assay medium incubated for (A) 0 min, (B) 30 min. (From Unnithan et al., 1984.)

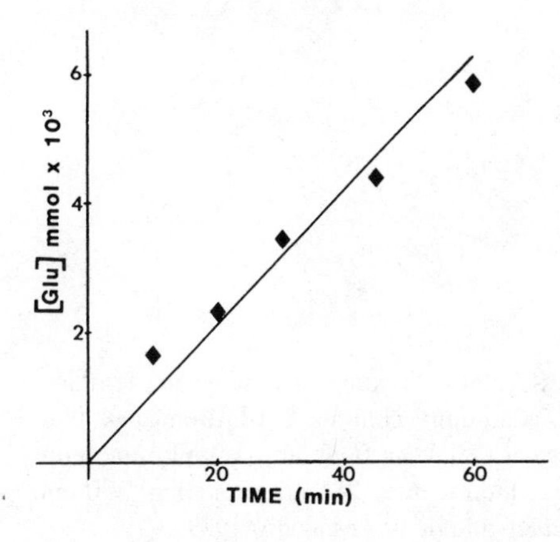

Figure 5.36 L-Glutamate formation with respect to incubation time at 37 °C due to the glutaminase activity. (From Unnithan et al., 1984.)

tail to the enzyme preparation. The solution was incubated at 37 °C, and at intervals the reaction was terminated by transferring samples to boiling sodium acetate buffer. The solutions were cooled and centrifuged, and the amino acids derivatized with *O*-phthaldialdehyde (OPA). Figure 5.35 illustrates the formation of reaction products by comparing the profiles obtained at zero time (A) and after 30 min (B) of incubation. The formation of the two reaction products, asparagine and glutamate, is clearly seen. Figure 5.36 shows the time course of glutamate formation.

The enzyme was prepared from bovine pancreas by previously published procedures.

5.4.4 Tryptophanase (Krstulovic and Matzura, 1979)

Tryptophanase catalyzes the conversion of tryptophan to indole and acetic acid. Pyridoxal phosphate is a required cofactor. The hplc method developed to assay this activity involves the separation of the tryptophan from the indole.

The separation was carried out by reverse-phase hplc (C-18 μBondapak) using a mobile phase of anhydrous methanol–water (1:1, v/v). The column was eluted isocratically and detected fluorometrically with excitation and emission wavelengths of 285 and 320 nm, respectively. The separation obtained is shown in Fig. 5.37.

The reaction mixture contained potassium dihydrogen phosphate buffered to pH 7.0 and bacterial cells. The cells were sonicated and preincu-

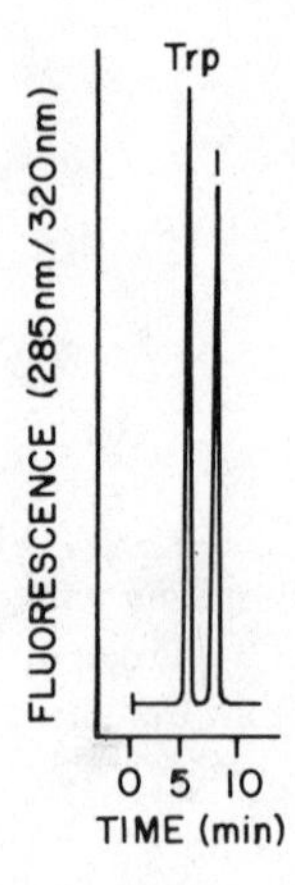

Figure 5.37 Separation of a synthetic mixture of tryptophan (Trp) and indole (I). Chromatographic conditions: column, C-18 μBondapak; eluent, anhydrous methanol–water (1:1 v/v); flow rate, 1.0 mL/min; temperature, ambient; detection, fluorescence, 285 nm excitation, 320 nm emission cutoff filter. (From Krstulovic and Matzura, 1979.)

bated with pyridoxal phosphate, and the reaction was started by the addition of tryptophan. At intervals, reactions were terminated with TCA and the solution clarified by centrifugation, filtered, and then analyzed. Several chromatograms showing the time-dependent increase in indole formation are illustrated in Fig. 5.38.

The tryptophanase was obtained from *Escherichia coli*.

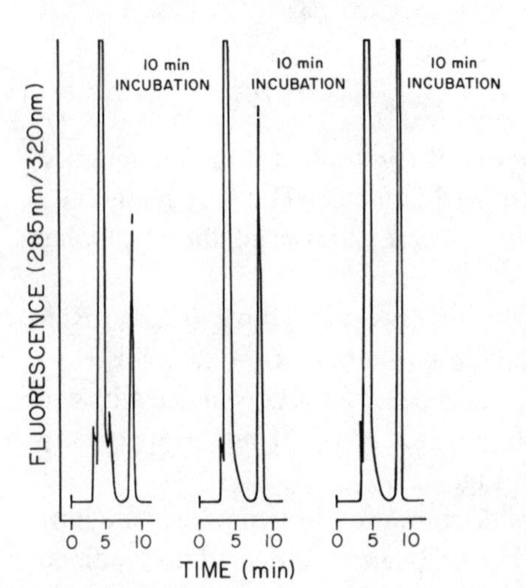

Figure 5.38 Chromatograms of the *E. coli* extracts incubated with tryptophan for 10, 20 and 40 min, respectively. Samples were deproteinated with TCA, and the extracts neutralized with solid tris (hydroxymethyl aminomethane). Chromatographic conditions same as in Fig. 5.37. Volume injected, 5 μl; attenuation, 0.5 μA. (From Krstulovic and Matzura, 1979.)

5.5 POLYAMINES

5.5.1 Ornithine Decarboxylase (Haraguchi et al., 1980)

Ornithine decarboxylase catalyzes the decarboxylation of ornithine to putrescine. The assay involves a preparation on CellexP, a conversion to the fluorescent derivative with fluorescamine, and separation on hplc.

The column was LiChrosorb RP-18, and the separation was carried out by elution with a gradient of 45–80% methanol and 0.1 *M* ammonium chloride in an acetate buffer (pH. 4.0).

The incubation mixture contained ornithine, pyridoxyl phosphate, and the enzyme. After incubation for 1 hr, the reaction was terminated with perchloric acid. The precipitate was removed by centrifugation, the supernatant extracted with chloroform–methanol (2:1), and the aqueous layer applied to a CellexP column. The putrescine was eluted, reacted with fluorescamine, and quantitated by hplc.

The enzyme was from rat intestinal mucosa and was partially purified by a 20–80% precipitation with ammonium sulfate.

5.5.2 Spermidine Synthetase (Porta et al., 1981)

The enzyme spermidine synthetase catalyzes the conversion of putrescine to spermidine. The separation of the substrate from the product was accomplished on a reverse-phase column (C-18) using a mobile phase of 60% methanol (v/v). The compounds in the eluent were detected by scintillation counting or by UV at 254 nm as shown in Fig. 5.39 (labeled Blank).

The assay was carried out in phosphate buffer with radioactive putrescine, decarboxylated *S*-adenosylmethionine, and enzyme. Reactions were incubated at 37 °C for 90 min, and the reactions were terminated by addition of perchloric acid. The solutions were clarified by centrifugation, and the polyamines were benzoylated and extracted and then analyzed. Figure 5.39 shows the analysis of samples removed at zero time (blank) and after 60 min incubation (sample) at 37 °C. The appearance of radioactive spermidine is shown. The rate of product formation is shown in Fig. 5.40.

The enzyme was prepared from mouse brain homogenized in phosphate buffer. An S-100 solution was prepared and used as the source of the synthetase.

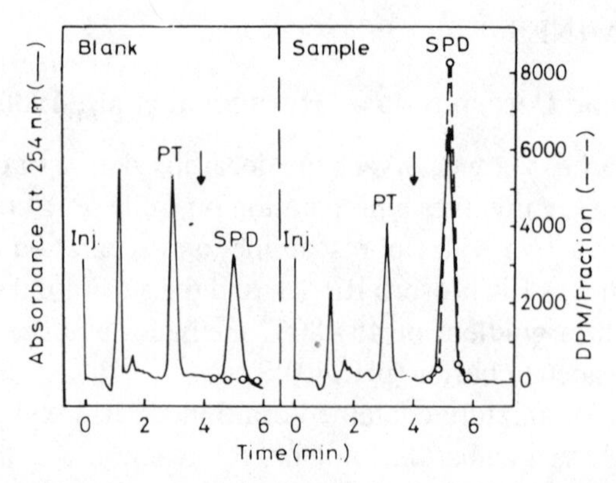

Figure 5.39 Hplc separation of the benzoyl derivatives of putrescine (PT) and spermidine (SPD) contained in a standard reaction sample and in the corresponding blank after 60 min of incubation at 37°C. Fractions (0.8 mL) were collected for determination of radioactivity. The arrows indicate a change in detector sensitivity from 0.2 to 0.05 absorbance units full scale. (From Porta et al., 1981.)

It is of interest to note the use of the "sensitivity-shift" procedures, described in Chapter Three, illustrated in Fig. 5.39. Arrows on the figure indicate where the detector sensitivity was increased to allow for the detection of levels of the product.

5.6 HEME BIOSYNTHESIS

5.6.1 δ-Aminolevulinic Acid Synthetase (Tikerpae et al., 1981)

This mitochondrial enzyme (ALA synthetase) catalyzes the formation of δ-aminolevulinic acid (ALA) from glycine and succinyl-CoA. This is the initial step in heme biosynthesis.

The assay incorporates a novel feature wherein radioactive succinyl-CoA is formed from succinate, which in turn is formed from radioactive δ-ketoglutarate. The succinyl-CoA then reacts with glycine to form ALA. For the assay, the ALA is converted to the pyrrole derivative 2-methyl-3-carbethoxy-4-(3-propionic acid) pyrrole.

The assay involves the isolation of the pyrrole by ion-paired reverse-phase hplc (Hypersil-SAS) with a mobile phase of methanol and water (45:155, v/v) and 0.005 mole/L 1-heptanesulfonic acid (PIC B-7). The radioactive product was detected by scintillation counting. The separation

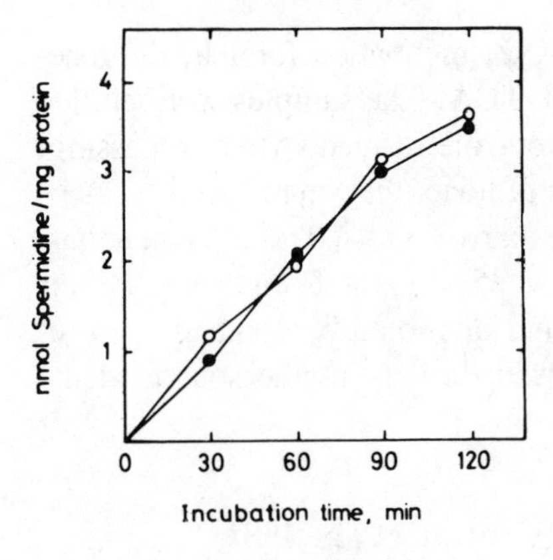

Figure 5.40 Mouse brain spermidine synthetase activity as a function of time. Both the absorbance (○) and the radiometric (●) determination of product formation represent the mean value of duplicate assays. The reproducibility was within 10%. (From Porta et al., 1981.)

obtained when the compounds were in distilled water is shown in Fig. 5.41(a). However, when the analysis was carried out on previously freeze-dried samples, the results shown in Fig. 5.41(b) were obtained due to the salts present.

The reaction mixture contained glycine, coenzyme A, buffer, α-ke-

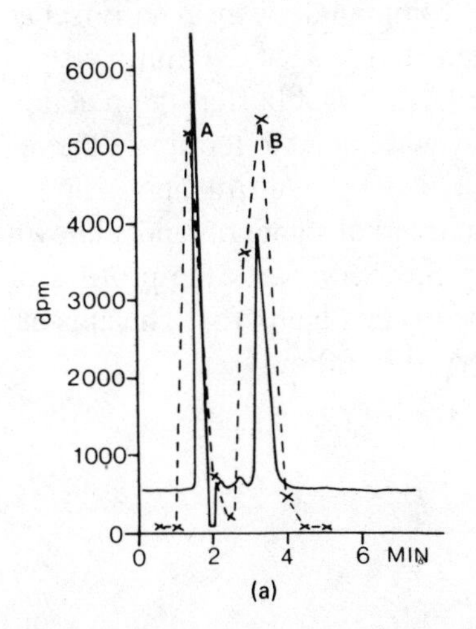

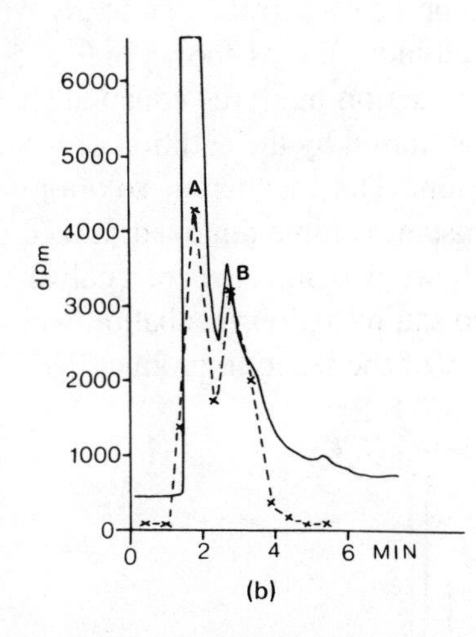

Figure 5.41 Separation of [5-^{14}C]α-ketoglutarate (A) and [4-^{14}C]aminolevulinic acid (B) by hplc on a Hypersil column with methanol–aqueous PIC B-7 (50:150, v/v) as the mobile phase at a flow rate of 1 mL/min. (a) Mixture in distilled water; (b) mixture in presence of salts recovered from a freeze-dried incubation mixture. (———) Spectrophotometric monitoring at 278 nm and 0.05 absorbance units full scale. (- - - -) Radioactive distribution (0.5 mL fractions counted). (From Tikerpae et al., 1981.)

toglutarate, and bone marrow lysate. After incubation for 1 h, the reaction was terminated by addition of 10% TCA. The samples were chilled and clarified by centrifugation, and the pyrrole formed. After processing, the pyrrole was isolated by hplc and its radioactivity quantitated.

The enzyme was obtained from bone marrow cells. The cells were harvested and pelleted by centrifugation at 2500 *g* for 5 min. The pellets were washed, resuspended, counted, and disrupted by sonication to release mitochondria. This lysate was used directly as the source of the enzyme.

5.6.2 5-Aminolevulinate Dehydrase (Crowne et al., 1981)

5-Aminolevulinic acid dehydrase (ALA dehydrase) is the second enzyme of the heme biosynthetic pathway. It catalyzes the condensations of two molecules of ALA to form porphyrobilinogen (PBG).

The assay involves the separation of ALA from PBG by ion-paired reverse-phase hplc (Hypersil SAS) with a mobile phase of methanol–water (22:78, v/v) and Pic B7 (0.005 *M* 1-heptanesulfonic acid) adjusted to pH 3.5. An internal standard of 2-methyl-3-carbmethoxy-4-(3-propionic acid)-pyrrole was used. All three compounds were readily separated in 6 min, as shown in Fig. 5.42. Detection was at 240 nm.

Reaction mixtures contained substrate (ALA) and buffer; the reaction was started by the addition of lysate. TCA was added to terminate the reaction. The incubation solution was clarified by centrifugation, and a constant volume removed, mixed with the internal standard, and a known volume was injected for analysis. Figure 5.43 compares the profiles of two samples after incubation with enzyme (a) and blank (b). The appearance of the reaction product PBG is observed in (a) only.

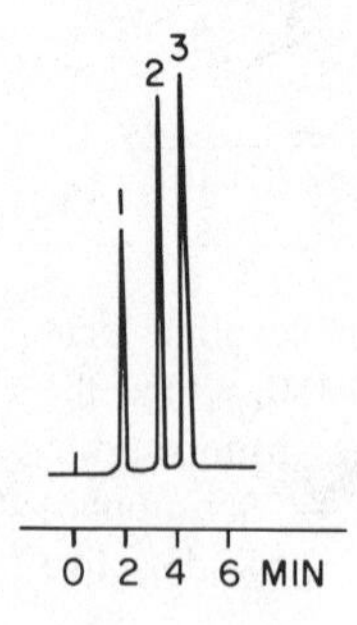

Figure 5.42 Separation of ALA (1), PBG (2), and internal standard (3). Stationary phase, Hypersil-SAS; mobile phase, methanol-PIC-B7 in water (22:78, v/v); pressure, 80 bar; flow rate, 1.2 mL/min; detection, 240 nm. (From Crowne et al., 1981.)

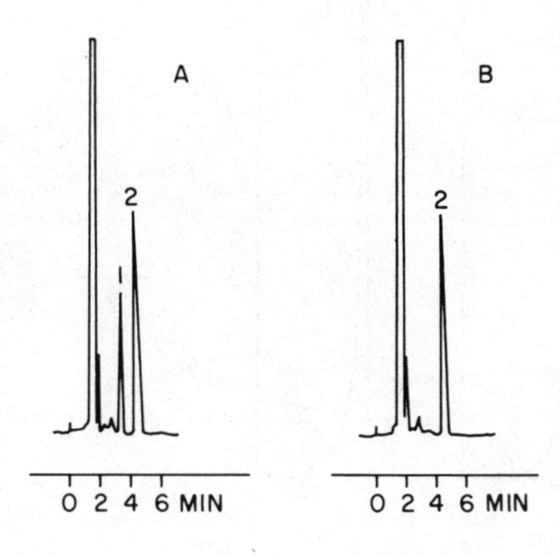

Figure 5.43 Separation of PBG (1) and internal standard (2) in incubation mixture. (A) Test; (B) blank. (From Crowne et al., 1981.)

Whole blood was hemolyzed in water, and the lysate was used as the enzyme solution.

5.7 CARBOHYDRATE METABOLISM

5.7.1 β-Galactosidase (Naoi and Yagi, 1981)

β-Galactosidase (β-D-galactoside galactohydrolase, EC 3.2.1.23) catalyzes the release of galactose from a variety of substrates including glycosphingolipids. The galactosidase specific for the release of terminal galactose from glycosphingolipids was studied using a derivative of a galactocerebroside, 1-*O*-galactosyl-2-*N*-DANS-sphingosine, as the substrate. The product of the β-galactosidase reaction, *N*-DANS-sphingosine, was measured using hplc.

The product was separated from the substrate on a normal phase silica gel column (Zorbax Sil) and eluted with methanol at 35°C as shown in Fig. 5.44 I and II. The concentration of the reactants was determined by the fluorescence intensity at 535 nm with an excitation wavelength at 340 nm.

The reaction mixture included the substrate, galactosyl-*N*-DANS-sphinogosine, suspended in citrate buffer and dispersed by sonication. The reaction was started by the addition of the substrate to the enzyme solution. The reaction mixture was incubated at 37°C for 30 min, and the reaction was terminated by the addition of a mixture of chloroform–

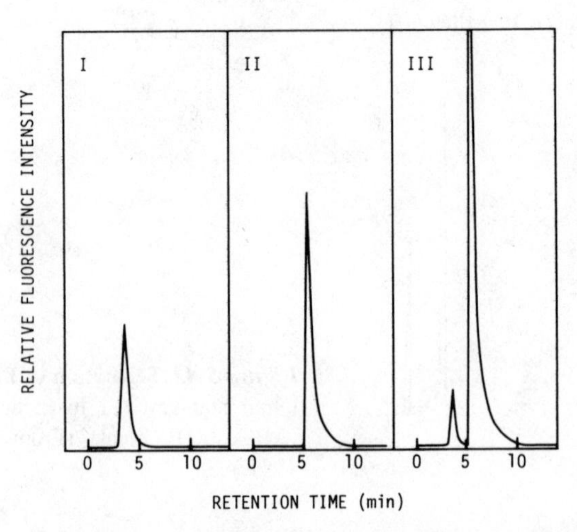

Figure 5.44 High-performance liquid chromatography of *N*-DANS-sphingosine and galacytosyl-*N*-DANS-sphingosine. Purified *N*-DANS-sphingosine (I) or galacytosyl-*N*-DANS-sphingosine (II). After reaction of crude β-galactosidase with galacytosyl-*N*-DANS-sphingosine, the sample was applied on hplc (III). (From Naoi and Yagi, 1981.)

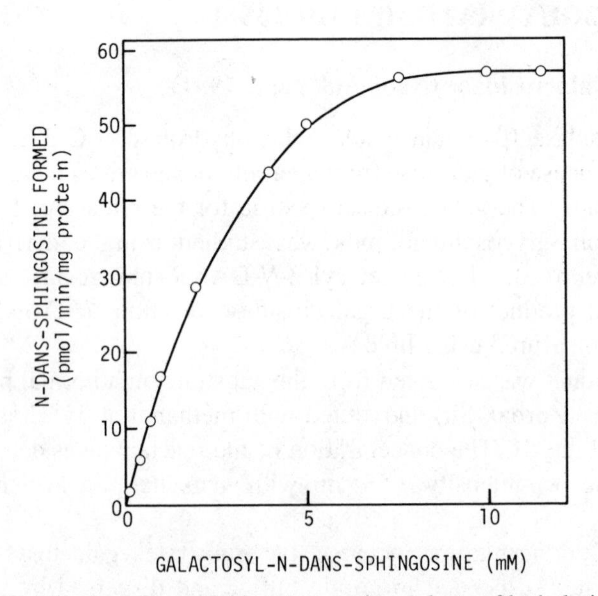

Figure 5.45 Relation between substrate concentration and rate of hydrolysis of galactosyl-*N*-DANS-sphingosine by crude β-galactosidase. Galactosyl-*N*-DANS-sphingosine was hydrolyzed by crude β-galactosidase (0.7 mg), and the formation of *N*-DANS-sphingosine was measured by hplc. (From Naoi and Yagi, 1981.)

methanol (2:1, v/v). After centrifugation the lower chloroform phase was recovered and evaporated to dryness, and the residue was dissolved in methanol and analyzed by hplc. Figure 5.44 III shows a chromatogram of a sample. The smaller of the two peaks, the product, is observed. Figure 5.45 shows the relationship between substrate concentration and rate of hydrolysis as determined by the hplc method.

The β-galactosidase was prepared from rat brain. The rat brain was homogenized in detergent and the homogenate centrifuged at 20,000 *g* for 20 min. The supernatant solution was made 60% saturated in ammonium sulfate. The precipitated protein was recovered, dissolved in detergent, and dialyzed against the same detergent. The dialysate was centrifuged, and the supernatant solution was used as the enzyme fraction.

5.7.2 Lactose-Lysine β-Galactosidase (Schreuder and Welling, 1983)

This particular β-galactosidase cleaves the compound lactose-lysine into β-galactose and fructose-lysine. This activity is of interest because it has been suggested as a possible marker for the presence of bacteria since it does not appear to be present in germ-free animals.

The reaction is followed by separation of the substrate, lactose-lysine, from the product, fructose-lysine, on a cation-exchange resin (Durrum DC6A) using an isocratic mobile phase of pyridine–acetic acid–water (6:60:176, v/v). OPA derivatives were formed and detected by fluorescence.

The reaction was carried out with ε-lactose-lysine as the substrate using a sodium phosphate buffer (pH 7.5). The reaction was started by the addition of the enzyme and after incubation for 1 h it was terminated with methanol containing β-alanine as an internal standard. Precipitated protein was removed by centrifugation, and samples of the supernatant solution were injected and analyzed by hplc. A chromatogram showing the analysis of a reaction mixture is given in Fig. 5.46.

The enzymes were obtained from mouse intestine. The intestinal segments were cut into pieces, homogenized with a glass rod, sonicated, and centrifuged. The supernatant solution was dialyzed and stored frozen for later use.

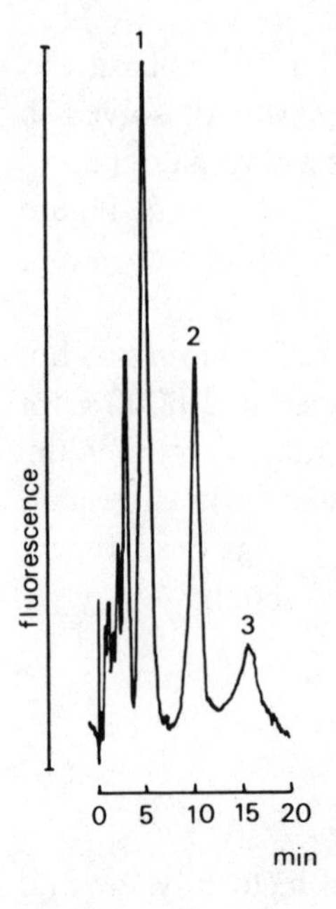

Figure 5.46 Chromatogram obtained after injection of a β-galactosidase incubation mixture onto a 45 × 3.6 mm cation-exchange column. Mobile phase: pyridine–acetic acid– water (6:60:176, v/v). Flow rate, 0.4 mL/min; temperature, 50 °C. In front of lactose-lysine some free amino acids elute which are the result of proteolytic activity in the intestinal enzyme preparation. (1) Lactose-lysine; (2) β-alanine (internal standard); (3) fructose-lysine. (From Schreuder and Welling, 1983.)

5.7.3 Arylsulfatase B (N-Acetylgalactosamine 4-Sulfatase) (Fluharty et al., 1982)

Arylsulfatase B catalyzes the hydrolysis of the sulfate from UDP-GalNAc-4-sulfate to form UDP-GalNAc and sulfate. This activity is found in normal fibroblasts but not in Maroteaux-Lamy fibroblasts.

The substrate was separated from the product using ion-paired reverse-phase hplc (Ultrasphere-ODS). The samples were eluted isocratically at room temperature using a mobile phase composed of methanol and 20 m*M* potassium phosphate (pH 2.5), (40:60, v/v) containing 15 m*M* tetrabutylammonium phosphate. The eluant was monitored at 262 nm.

The reaction was carried out in a solution containing UDP-GalNAc-4-S, acetate buffer at pH 3.5, and the arylsulfatase B. Reactions were incubated at 37 °C for 30 min and terminated by heating in a boiling water bath for 1 min. Longer heating resulted in the formation of a new component which overlapped the product peak on the hplc. The heated material was centrifuged at 8000 *g* for 2 min, and a sample was used directly for analysis. Figure 5.47 shows chromatograms of samples taken at zero time and after 30 min incubation. The appearance of the new peak at about 4.5

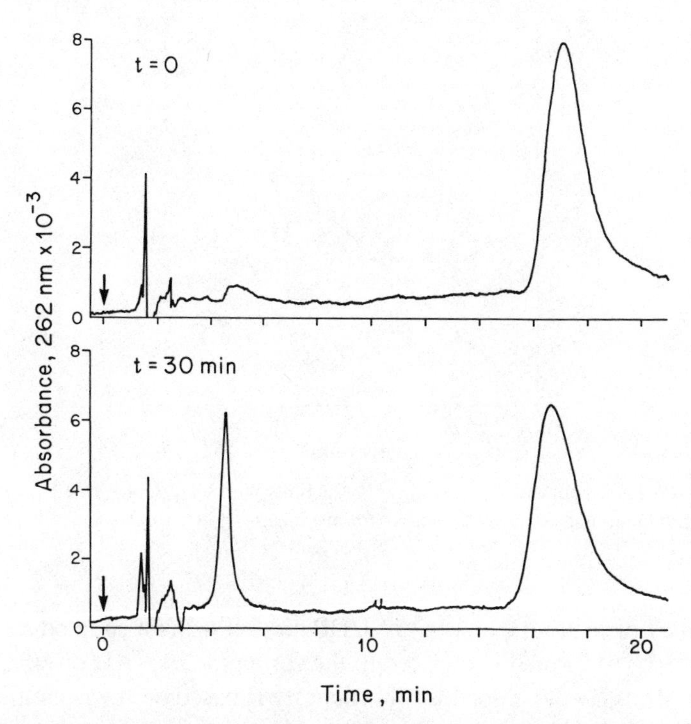

Figure 5.47 Hplc analysis of arylsulfatase B reaction with UDP-GalNAc-4-S. The reaction mixture contained 5 mU of enzyme. (From Fluharty et al., 1982.)

min indicates the formation of the reaction product. The time course of sulfatase activity is shown in Fig. 5.48.

The enzyme was prepared from human liver. This substrate is not hydrolyzed by arylsulfatase A.

5.7.4 Galactosyltransferase (Hymes and Mullinax, 1984)

Galactosyltransferase catalyzes the transfer of galactose from UDP-galactose to either glucose or *N*-acetylgalactosamine or to N-acetylglucosamine when this is a terminal residue of complex oligosaccharides. The activity occurs in many different types of tissues and body fluids.

Most assays for this activity are of one of two types: either they determine the amount of radioactive product formed from radioactive UDP-

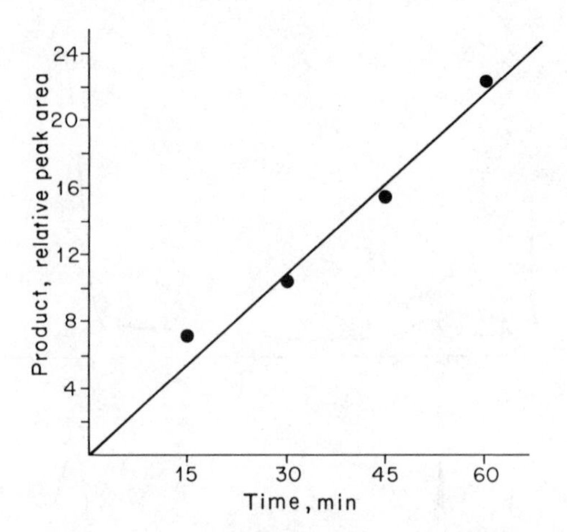

Figure 5.48 Time course of arylsulfatase B reactions with UDP-GalNAc-4-S. The reaction mixture contained 3 mU of enzyme. (From Fluharty et al., 1982.)

Gal or they measure the amount of UDP present using a coupled assay in which the UDP formed is coupled to the formation of NAD or NADH.

The hplc assay developed for this activity is based on the isocratic separation of UDP, UDP-Gal, and UMP on a amino-bonded column (μBondapak NH_2 column) eluted with 0.167 *m* KH_2PO_4 (pH 4.0). The separation is shown in Fig. 5.49. Detection was at 260 nm.

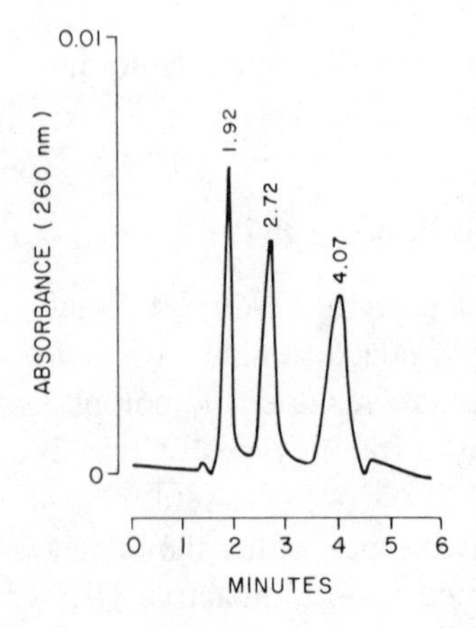

Figure 5.49 Isocratic hplc separation and UV absorption detection of UMP, UDP-Gal, and UDP as sequentially eluted. A mixture containing 500 pmol of each compound was injected. (From Hymes and Mullinax, 1984.)

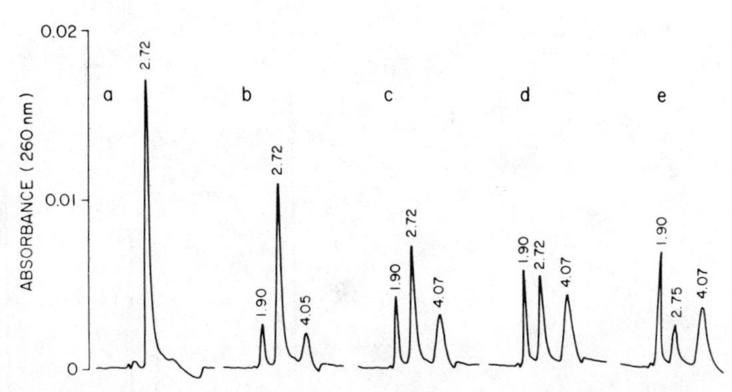

Figure 5.50 Conversion of UDP-galactose to UMP and UDP. Aliquots of galactosyltransferase reaction (glucose substrate, bovine milk enzyme) were removed at 0, 15, 30, 45, and 60 min and assayed by hplc with UV detection (a–e, respectively). Peaks representing UMP, UDP-Gal, and UDP are detected in later samples. (From Hymes and Mullinax, 1984.)

The reaction mixture contained *N*-acetylglucosamine, UDP-Gal, $MnCl_2$, and buffer at pH 8.0. The reaction was started by the addition of the enzyme. Samples were transferred at intervals to cacodylate buffer on ice (pH 6.5) to terminate the reaction. Samples (10 μL) were analyzed by hplc. The conversion of UDP-Gal to UDP is shown in Fig. 5.50. Each panel, labeled a–e, represents a different time point, from 0 to 60 min. During the incubation, the disappearance of the substrate and the formation of the two products is seen.

The enzyme was obtained from commercial sources and human serum.

5.7.5 Uridine Diphosphate Glucuronosyltransferase (Matsui and Nagai, 1980; To and Wells, 1984)

Uridine diphosphate glucuronosyltransferase catalyzes the transfer of glucuronic acid from uridine diphosphate glucuronic acid (UDPGA) to various substrates such as 4-nitrophenol (4-NP), phenolphthalein (P), or testosterone.

The substrates and the product were separated by reverse-phase hplc (styrene–divinylbenzene copolymers) at 40°C with a methanol–water (65%, v/v) mobile phase containing 0.01 *N* HCl as shown in Fig. 5.51. Absorbance at 300 nm was measured, and quantitation was carried out from peak height measurements.

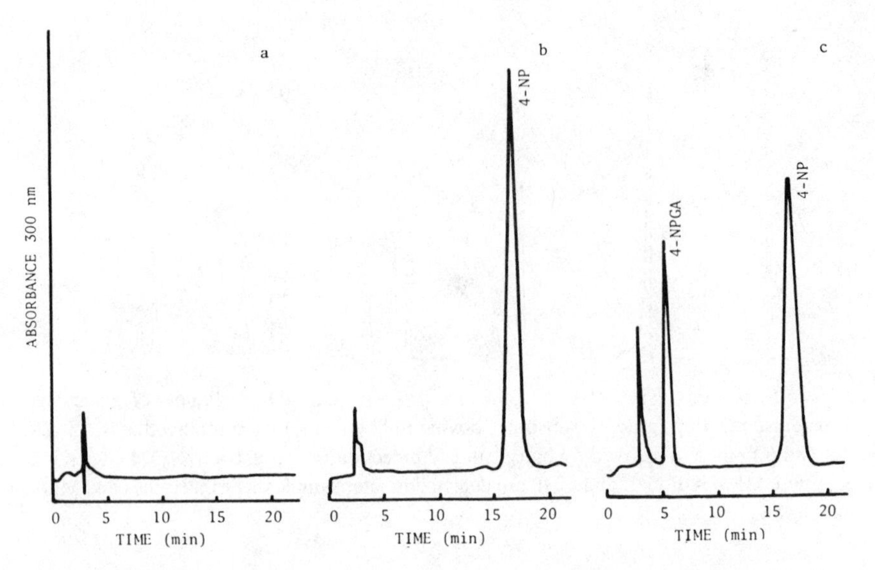

Figure 5.51 Chromatograms recorded in the assay of UDPGT toward 4-NP. (a) Assay without 4-NP, (b) assay without UDPGA, and (c) assay with all ingredients. Chromatographic conditions: column, Hitachi gel No. 3010, 3 × 500 mm, mobile phase, 0.01 *N* HCl in 65% (v/v) methanol; detection, 300 nm; flow rate, 0.8 mL/min; pressure, 60 kg/cm^2; column temperature, 40°C. (From Matsui and Nagai, 1980.)

The assay mixture contained, in a final volume of 1 mL, 0.4–1.2 mg of the microsomal fraction protein, UDPGA as donor and either 4-NP or P as the acceptor. The reactions were terminated in a boiling water bath (1–5 min). Methanol was added, and the solution was clarified by centrifugation. Samples of the supernatant solution were analyzed directly. Figure 5.51 shows three chromatograms obtained from samples of reaction mixtures without substrates (a and b) and the complete reaction mixture (c). The formation of product is clearly seen.

The enzyme was prepared from rat liver. An homogenate was prepared in sucrose with a Teflon–glass homogenizer. A series of differential centrifugations (2000 *g* for 10 min, 16,000 *g* for 45 min, and 105,000 *g* for 60 min) produced a microsomal pellet that was used as the enzyme source.

In another study, the glucuronic acid was transferred from UDPGA to the acceptor, α-naphthol.

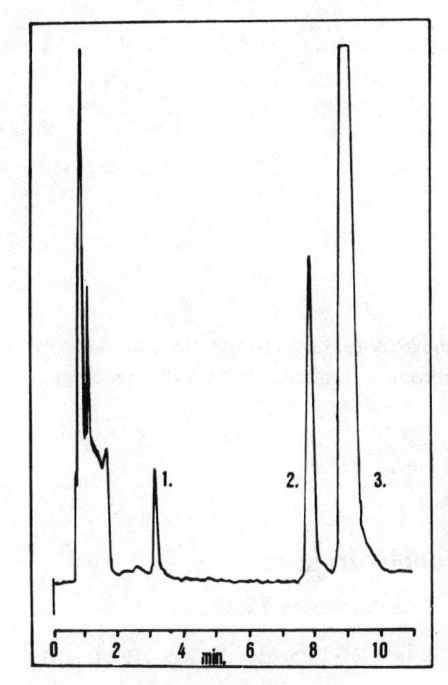

Figure 5.52 Chromatogram of the hplc resolution of α-naphthol glucuronide. Peak identification: (1) α-naphthol glucuronide; (2) β-naphthol, the internal standard; (3) α-naphthol, the substrate. Solvent: 0.1 *M* acetic acid–methanol (55:45, v/v); flow rate 1.5 mL/min; wavelength, 240 nm. (From To and Wells, 1984.)

In this assay the substrate α-naphthol was separated from the reaction product α-naphthol glucuronide by reverse-phase hplc (C-18 column) with a solvent of 0.1 *M* acetic acid–methanol (55:45, v/v). Absorbance was monitored at 240 nm.

The reaction mixture contained the substrate, α-naphthol (5.0 m*M*), in Tris-HCl buffer (pH 7.4) with 10 m*M* magnesium chloride and dimethylsulfoxide. The second substrate, UDPGA, was added, and the reaction mixture was preincubated for 5 min. The microsomal protein was then added as the source of enzyme activity. Incubations were at 37°C for up to 20 min. The reactions were terminated by the addition of ice-cold methanol, and any insoluble material was removed by centrifugation. The supernatant solution was dried under nitrogen, redissolved in methanol, and injected onto the hplc column for analysis. Figure 5.52 shows a chromatogram illustrating the separation of the substrate α-naphthol and the reaction product, glucuronide. Figure 5.53 shows the rate of glucuronide formation during the incubation.

The enzyme was obtained from mice. Hepatic microsomal fractions were used for these studies.

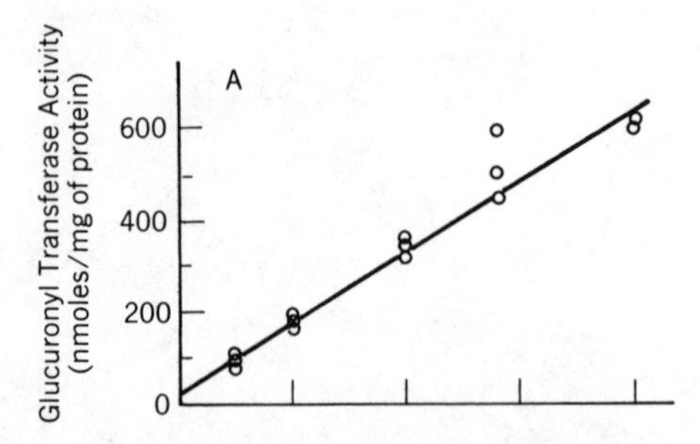

Figure 5.53 Effect of incubation time on enzyme activity. Glucuronyltransferase activity was measured by the amount of α-naphthol glucuronide produced after varying incubation time periods. (From To and Wells, 1984.)

5.7.6 α-Amylase (1,4-α-D-Glucaglucanohydrolase, EC 3.2.2.1) (Haegele et al., 1981)

α-Amylase is a hydrolase that cleaves 1,4-α glycosidic bonds and is important in the diagnosis of pancreatitis.

This assay involves the use of maltoheptose (seven glucose residues in a linear 1,4-α linkage) as the substrate. The activity degrades the substrate to smaller oligosaccharides, which are then subjected to α-glucosidase treatment to generate glucose. The α-glucosidase and the hexokinase are used as "indicator reactions."

The assay relies on the separation of maltose, maltotriose, maltotetrose, maltopentose, maltohexose, and maltoheptose. These compounds were separated by partition chromatography on a cation-exchange column (Nucleosil 10 SA) with a mobile phase of acetonitrile–water (72.5:27.5) eluted isocratically. The detector was a differential refractometer.

The enzyme assays were carried out in a phosphate buffer (pH 7.0), at 30°C with maltoheptose as the substrate. At intervals after the addition of the substrate to a solution of the amylase and indicator enzymes, samples were removed and diluted into a slurry of a mixed-bed ion exchanger to stop the reaction. The oligosaccharides were recovered and concentrated by lyophilization. The samples were dissolved in acetonitrile–water (1:1, v/v) and injected onto the hplc column for analysis. The results of the assay are shown in Fig. 5.54. The amount of glucose (peak 1), is seen to in-

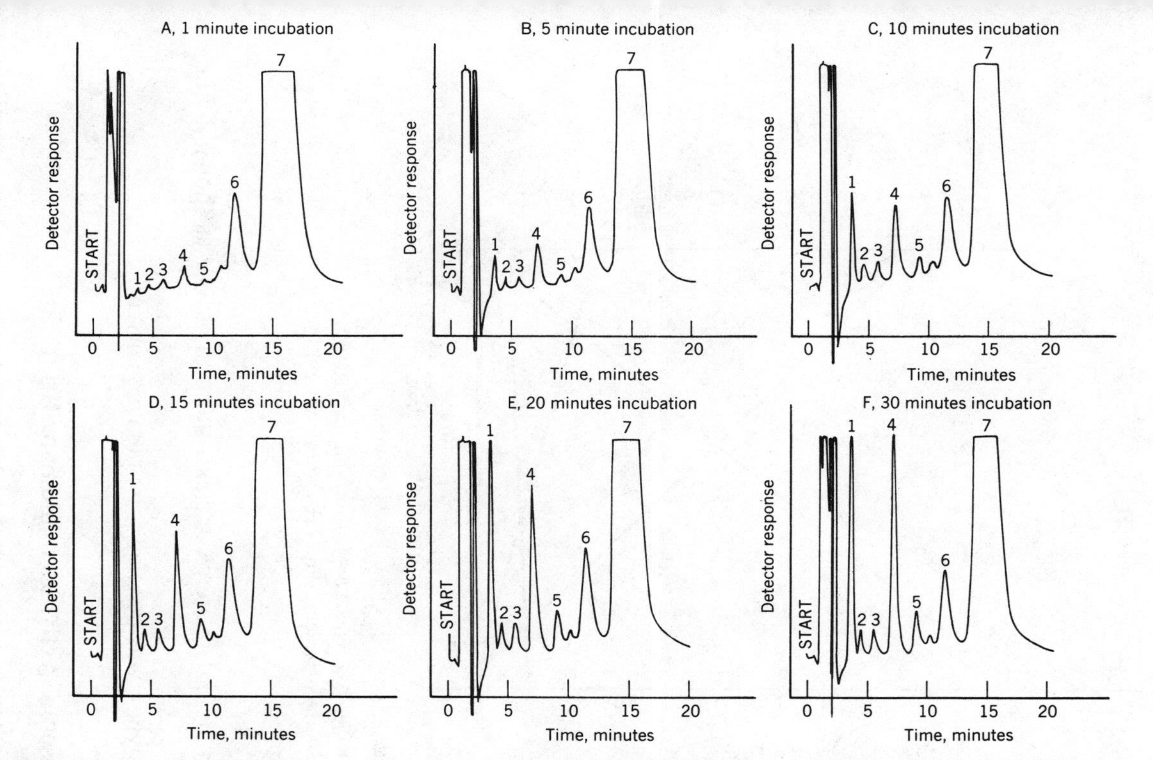

Figure 5.54 Chromatograms of the action patterns of maltoheptaose after the indicated periods of incubation with α-amylase and α-glucosidase. Peaks: (1) glucose; (2) maltose; (3) maltotriose; (4) maltotetraose; (5) maltopentaose; (x) compound A; (6) maltohexaose; (7) maltoheptaose. (A) Pure maltoheptaose as used for the assay. (B) Blank sample before the addition of substrate. (C–H) Chromatograms after 1, 5, 10, 15, 20, and 30 min, respectively, of incubation. Chromatographic conditions: column, 10 μm Nucleosil SA (250 × 4 mm); solvent, acetonitrile–water (72.5:27.5); flow rate, 0.7 mL/min; temperature, 27°C; detection, differential refractometer, full scale = 2×10^{-6} refractive index units. (From Haegel et al., 1981.)

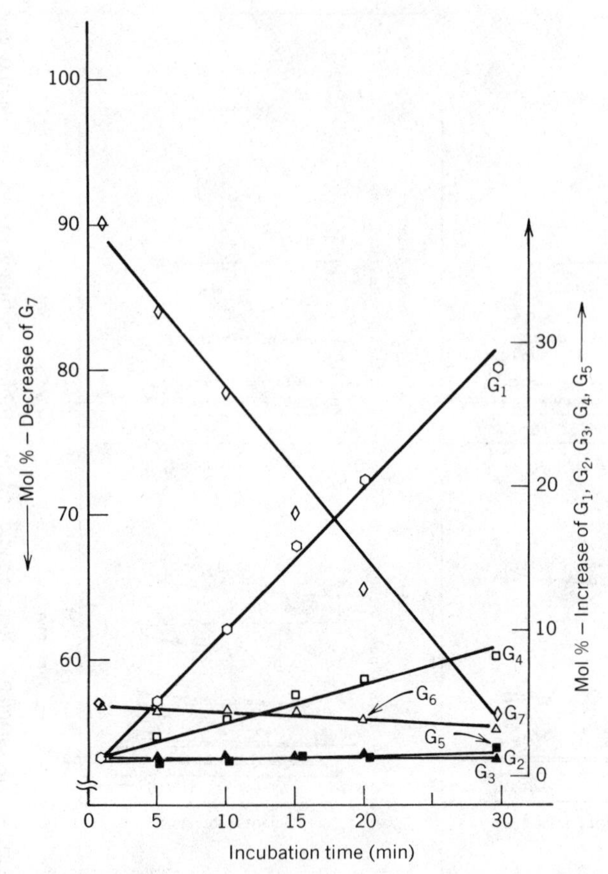

Figure 5.55 Kinetics of the coupled enzyme system α-amylase–α-glucosidase with maltoheptaose in mole percent. (From Haegele et al., 1981.)

crease when these data are plotted as a function of time as shown in Fig. 5.55.

The α-amylase was purified from human pancreas.

5.7.7 Lysosomal Activities (Sandman, 1983)

The hplc method has been used to assay a number of activities usually found associated with lysosomal vesicles. All these assays utilize the fluorometric compound 4-methylumbelliferone (4-MU). When carbohydrates, lipids, phosphates, or sulfates are conjugated with 4-MU, these compounds can be used as substrates for glycosidases, lipases, acid phosphatases, and arylsulfatases. The activity is determined by the release of 4-MU.

The separation of free 4-MU (the reaction product) from the conjugate (the substrate) was carried out by reverse-phase hplc on a PRP-1 column (styrene–divinylbenzene copolymer). A guard column was also used. The column was eluted isocratically for all the assays except the sulfatases, with 0.04 *M* glycine–sodium hydroxide buffer (pH 10.32), in aqueous methanol. For the sulfatase activity separations, the glycine buffer with 20% methanol was used as mobile phase. The fluorescence of the eluent was monitored with excitation and emission wavelengths of 360 and 455 nm, respectively.

The conditions for enzymatic analysis varied with the enzyme under study. In general, the reactions were started by the addition of the enzyme activity, and incubations were at 37°C. The reaction was stopped by the addition of 100% methanol, and precipitated protein was removed by centrifugation. Samples of the supernatant solution were analyzed by hplc. A chromatogram of a glycosidase activity is shown in Fig. 5.56. The peak of free 4-MU is noted.

The enzyme was obtained from urine samples.

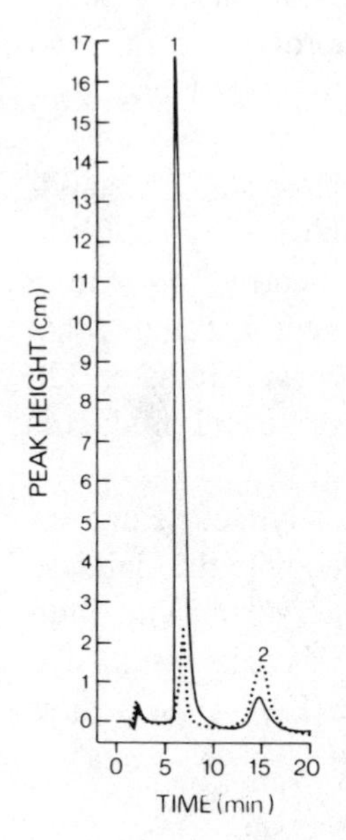

Figure 5.56 Chromatogram of urinary β-N-acetyl-D-glucosaminidase (NAG) assay reaction mixture (solid curve) and assay blank (dotted curve). Peaks: (1) 4 MU, (2) 4 MU-NAG. Enzyme source: Tenfold diluted urine. Detector sensitivity for the assay blank was twice that of the assay reaction mixture. (From Sandman, 1983.)

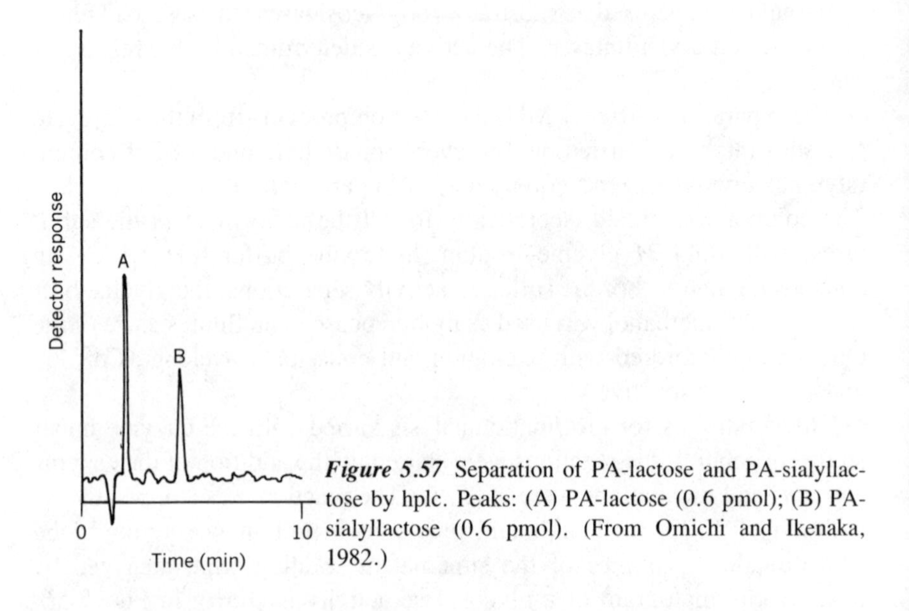

Figure 5.57 Separation of PA-lactose and PA-sialyllactose by hplc. Peaks: (A) PA-lactose (0.6 pmol); (B) PA-sialyllactose (0.6 pmol). (From Omichi and Ikenaka, 1982.)

5.7.8 Sialidase (Omichi and Ikenaka, 1982)

Sialidases, which are found in microorganisms and animal tissue, cleave sialic acid residues from terminal saccharide residues of carbohydrates. An assay was developed for this activity using the fluorogenic compound α-D-*N*-acetylneuraminyl-β-D-(2→3)galactopyranosyl-(1→4)-1-deoxy-1-[(2-pyridyl)-aminol]-D-glucitol (PA-sialyllactose).

The assay method involves the separation of the substrate, PA-sialyllactose, from the product, PA-lactose. This is accomplished by gel filtration hplc on a TSK-Gel LS 410 column. Elution was carried out with a mobile phase of 0.1 *M* acetic acid. The compounds were detected with a fluorescence detector using excitation and emission wavelengths of 320 and 400 nm, respectively. The separation of the substrate and product is shown in Fig. 5.57.

The reaction mixture contained, in 15 μL, the PA-sialyllactose in a sodium acetate buffer at pH 5.0. The reaction was started with the sialidase preparation and incubated at 37°C. At appropriate intervals, 20 μL samples were removed and injected directly for hplc analysis. The results of a reaction are shown in Fig. 5.58. The appearance of the PA-lactose peak is an indication of sialidase activity.

The activity was obtained from urine.

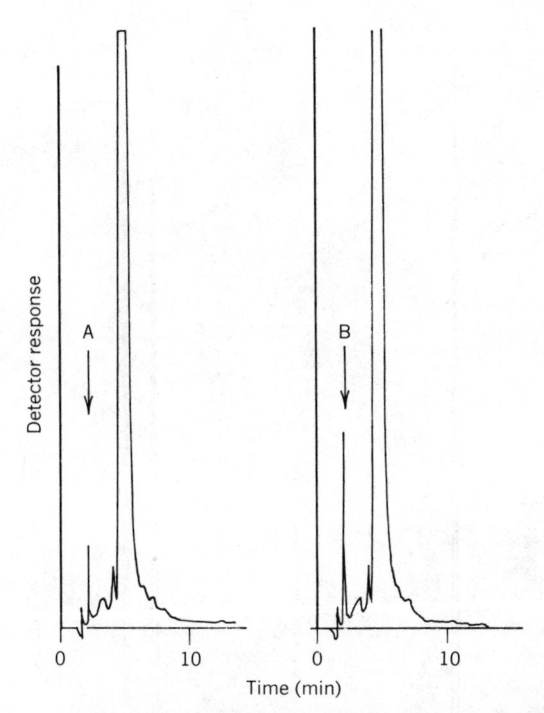

Figure 5.58 Chromatogram of the digest of PA-sialyllactose by urine. PA-sialyllactose (0.167 m*M*) was incubated with dialyzed urine in 0.1 *M* sodium acetate buffer (pH 5.0) at 37°C for 2 h. (A) substrate + urine heated at 100°C for 10 min (control). (B) Substrate + urine. The arrows show the elution position of PA-lactose. (From Omichi and Ikenaka, 1982.)

5.7.9 Cytidine Monophosphate–Sialic Acid Synthetase (Petrie and Korytnyk, 1983)

Cytidine monophosphate–sialic acid synthetase (EC 2.7.7.43) catalyzes the activation of *N*-acetylneuraminic acid (NANA) by CTP to form CMP-NANA. The hplc assay method developed for this activity involves the separation of the three compounds NANA, CTP, and CMP-NANA. However, at the wavelength used for detection, only the last two are detected.

In this assay, separation is carried out by ion-paired reverse-phase hplc using an Ultrasphere-ODS-IP column with gradient elution from 5% acetonitrile, in a buffer solution (pH 7.5) of 5 m*M* tetrabutylammonium phosphate and 5 m*M* sodium phosphate, to 10% acetonitrile in the same

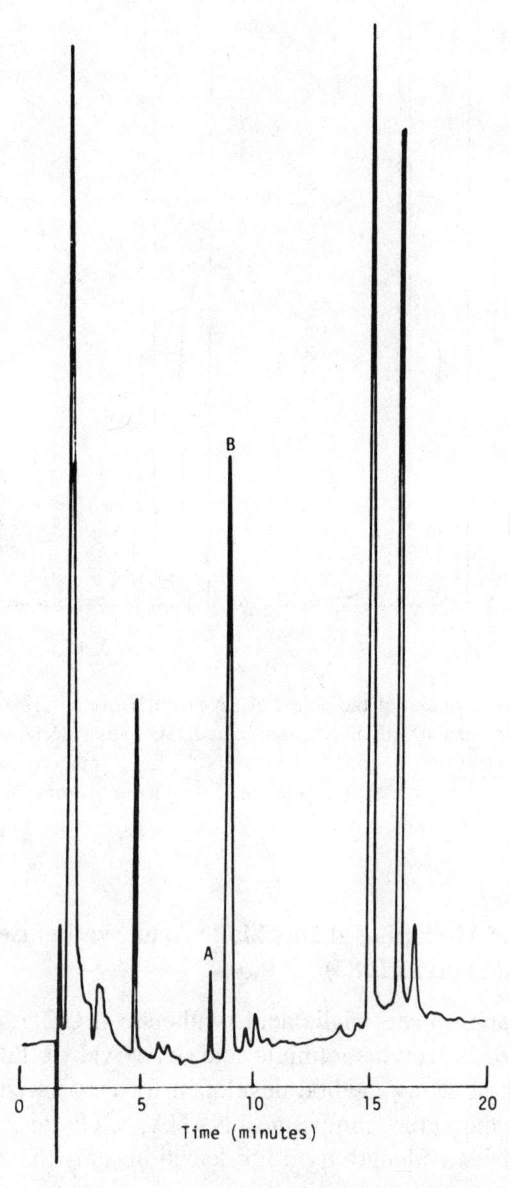

Figure 5.59 Hplc chromatogram of a typical enzyme assay sample with NANA and CTP as substrates. Peaks: (A) CMP, (B) CMP-NANA. (From Petrie and Korytnyk, 1983.)

buffer. Detection of CMP and CMP-NANA was at 270 nm. The separation obtained is shown in Fig. 5.59.

The reaction mixture contained CTP, glutathione, $MgCl_2$, and Tris-HCl at pH 9.0 into which the lyophilized enzyme was homogenized immediately before the assay. The reaction was started by the addition of NANA to the enzyme solution. Incubations were for 30 min and were terminated by the addition of cold acetonitrile. The solutions were clarified by centrifugation and by passage through small columns (ODS). The eluant was analyzed by hplc.

The synthetase was prepared from calf brains by homogenization, centrifugation (1500 *g* for 20 min), and lyophilization of the supernatant solution.

5.8 STEROID METABOLISM

5.8.1 Δ^5-3β-Hydroxysteroid Dehydrogenase (Suzuki et al., 1980)

The activity Δ^5-3β-hydroxysteroid dehydrogenase catalyzes the conversion of pregnenolone to progesterone, the progestational hormone of the placenta and corpus luteum. The product has an absorption maximum at 240 nm, and therefore detection can be readily carried out by UV absorbances near this wavelength.

In this assay the amount of product formed during the reaction was determined on a reverse-phase hplc (μBondapak) column containing cyanopropylsilane as the functional group. The column was eluted isocratically with a mixture of acetonitrile and water (1:4, v/v). Detection was at 254 nm.

The substrate, pregnenolone (158 nmol) dissolved in propylene glycol, was added to the incubation flask containing the enzyme preparation and NAD in a final volume of 5 mL. After an incubation period of 60 min, the reaction was terminated by addition of 15 μL of dichloromethane, and radioactive progesterone was added as a recovery standard. The organic phase was recovered, dried, and redissolved in 70% ethanol, and a sample was analyzed by hplc. Figure 5.60 shows an analysis of an incubation mixture with and without incubation with the ovarian preparation. The substrate (not seen) is converted exclusively to progesterone.

The ovaries of rats were used in the preparation of the active fraction.

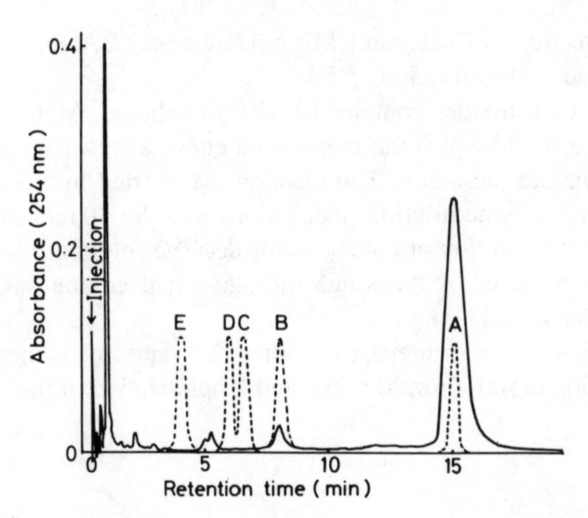

Figure 5.60 Hplc chromatograms of steroids. The elution profiles of five standard Δ^4-3-oxosteroids are illustrated as the broken line. (A) Progesterone; (B) 17α-hydroxyprogesterone; (C) androstenedione; (D) testosterone; (E) 11-deoxycortisol. The solid line is a chromatogram of a defatted extract obtained by incubation of pregnenolone with ovarian homogenates. (From Suzuki et al., 1980.)

5.8.2 11-β-Hydroxylase and 18-Hydroxylase (Gallant et al., 1978)

The 11-β-hydroxylase found in the adrenal cortex catalyzes the hydroxylation of 11-deoxycorticosterone to corticosterone. The enzyme requires NAD as a cofactor and contains heme as the prosthetic group.

The substrate, 11-deoxycorticosterone, was separated from the two reaction products, corticosterone (11-β-hydroxylation) and 18-hydroxyl-11-deoxycorticosterone (18-hydroxylation), on reverse-phase hplc (MicroPak silica) with a mobile phase of 16% tetrahydrofuran in H_2O. Figure 5.61A shows a chromatogram of the separation of the authentic steroids.

Hydroxylase activity was determined in a reaction mixture containing mitochondrial protein and 11-deoxycorticosterone (60 μ*M*). The reaction was started by the addition of 10 m*M* isocitrate as the source of reducing equivalents. At intervals during the incubation, samples were removed and placed into methylene chloride containing 11-deoxycortisol as an internal standard and extracted. After further processing of the extracts, the samples were injected, and the amount of 11-β- and 18-hydroxylation re-

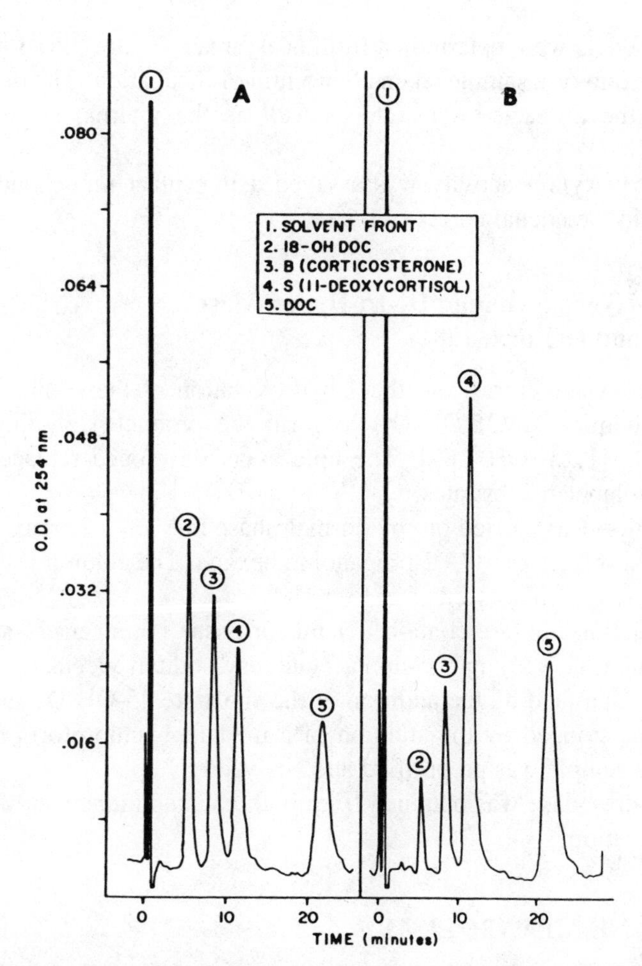

Figure 5.61 (A) Hplc chromatogram obtained with 10 μL of a mixture of authentic steroid standards (1.0 μg of each) on a Varian 0.2 × 25 cm MicroPak silica column with 16% tetrahydrofuran in water. The flow rate was 60 mL/h at 300 atm. (B) A typical chromatographic profile obtained with extracts of rat adrenal mitochondria incubated with 60 μM 11-deoxycorticosterone. An aliquot (0.8 mL) was removed at 5 min of incubation and extracted with 10 mL of methylene chloride containing 10 μg of internal standard (11-deoxycortisol). The extract was evaporated to dryness and solubilized with 50 μL of ethanol. Ten microliters was then injected onto the column. (From Gallant et al., 1978.)

action products were determined from peak areas. Figure 5.61B shows a chromatogram of a sample taken after 5 min of incubation. The two products and the unreacted substrate as well as the internal standard are shown.

The hydroxylase activity was assayed using intact mitochondria obtained from the adenal cortex of the rat.

5.8.3 25-Hydroxyvitamin D_3-1α-Hydroxylase (Tanaka and DeLuca, 1981)

This hydroxylase carries out the 1-hydroxylation of the compound 25-hydroxyvitamin D_3 (25-OH-D_3) to form the product 1,25-dihydroxyvitamin D_3 [1,25-$(OH)_2$-D_3]. The hplc assay developed replaces those using radiolabeled substrates.

Separation was carried out by normal phase hplc on a Zorbax Sil column with a solvent of 10% 2-propanol in hexane. The column was eluted isocratically and monitored at 254 nm.

The reaction mixture contained 1 mL of tissue homogenate, sucrose, Tris-acetate (pH 7.4), magnesium acetate, and sodium succinate. The reaction was initiated by the addition of the substrate 25-OH-D_3 and at intervals was stopped by the addition of a methanol–chloroform mixture (2:1). The sample was prepurified and analyzed by hplc.

The hydroxylase was obtained from both chicken kidney and liver by homogenization.

5.9 PURINE METABOLISM

5.9.1 Nicotinate Phosphoribosyltransferase (Hanna and Sloan, 1980)

Nicotinate phosphoribosyltransferase catalyzes the formation of nicotinate mononucleotide (NMN) and pyrophosphate from 5-phosphoribosyl-α-D-pyrophosphate (PRibPP) and free nicotinic acid. The reaction requires the hydrolysis of ATP to ADP.

In this study, the reactants, ATP, NMN, and ADP, were separated by reverse-phase hplc (μBondapak C-18) eluted isocratically with a mobile phase of 25 m*M* $(NH_4)_3PO_4$ (pH 8.0). Figure 5.62 illustrates the separation obtained under these conditions.

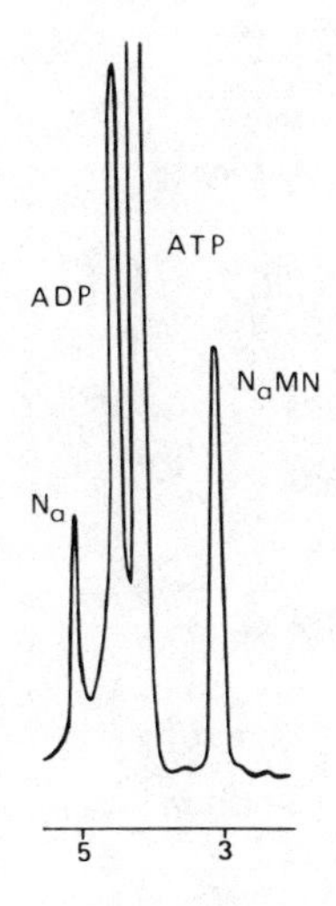

Figure 5.62 Elutions of a stock solution of ATP, ADP, nicotinate, and N_aMN through a μBondapak C-18 column using 25 m*M* $(NH_4)PO_4$ at pH of 8.0. Stock solutions of each reactant were used to assign the peaks. Elution conditions: 5 μL sample injection volumes, 0.7 mL/min flow rate, 25 °C. (From Hanna and Sloan, 1980.)

The assay mixture contained in a final volume of 5 mL $MgCl_2$, nicotinate, PRibPP, ATP, and Tris-HCl (pH 8.0). The reaction was started by the addition of enzyme (100 μg). At suitable intervals, 0.5 mL samples were removed and placed in a boiling water bath to terminate the reaction, then centrifuged and the supernatant solutions filtered. Volumes of 5 μL were removed and analyzed. Figure 5.63 shows a chromatogram of samples removed at various times during the incubation. The formation of the two products, ADP and NMN, is clearly observed.

The enzyme was purified from yeast and was free of NMN adenyltransferase, ATPase, and adenylate kinase activities.

5.9.2 5′-Nucleotidase (Sakai et al., 1982)

The enzyme 5′-nucleotidase is a phosphomonoesterase that catalyzes the dephosphorylation of 5′-monophosphates of purines and pyrimidimes to the nucleoside and free phosphate.

The assay requires the separation of the nucleoside monophosphate (the substrate) from the nucleoside (the product). This can be readily accomplished by ion-pair reverse-phase hplc with a mobile phase of 5% methanol–5 m*M* potassium dihydrogen phosphate, and 0.25 m*M* 1-decanesulfonic acid was also added to the mobile phase. The elution was carried out at room temperature and the eluent monitored at 254 nm.

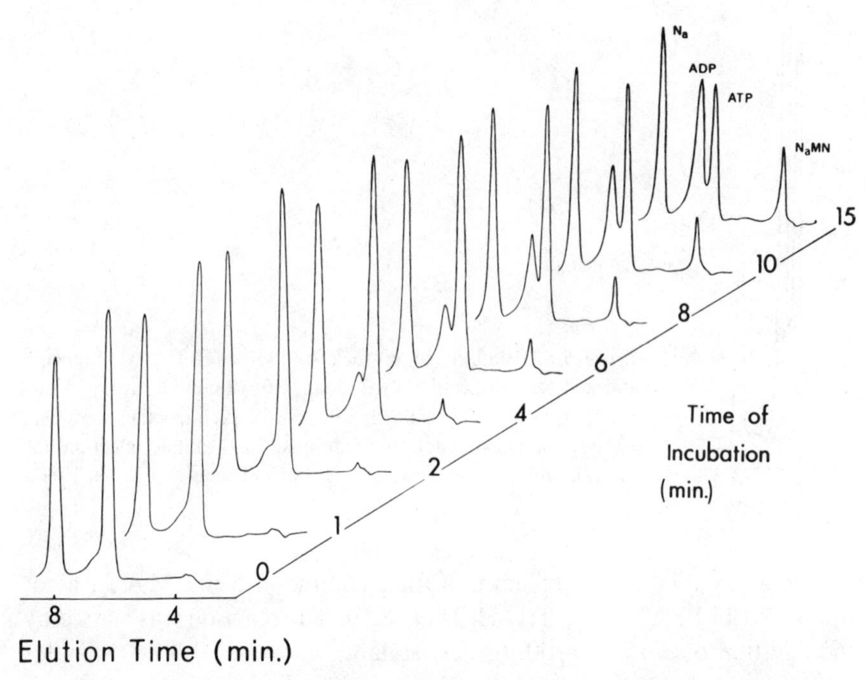

Figure 5.63 Elutions of the nicotinate phosphoribosyltransferase (N-PRTase) assay solution through a μBondapak C-18 column after various enzyme incubation times. The incubation mixture contained 5 m*M* $MgCl_2$, 100 μ*M* nicotinate, 75 μ*M* ATP, 30 μ*M* PRibPP, and 25 μL of 4 mg/mL N-PRTase in 50 m*M* Tris-HCl (pH 8). Elution conditions: 5 μL sample injection volumes, 0.7 mL/min flow rate, 25 m*M* $(NH_4)PO_4$ (pH 8) elution buffer, 25°C. (From Hanna and Sloan, 1980.)

The reaction mixture contained Tris-HCl-buffered UMP (purified free of uridine by ion-exchange chromatography) as the substrate and $MgCl_2$. The reaction was started by the addition of the enzyme, and the incubation was carried out at 37°C for 60 min. The reaction was terminated by placing the reaction tubes in a boiling water bath for 3 min. After dilution and centrifugation, the supernatant solution was analyzed by hplc.

Figure 5.64 shows the separation of UMP and uridine with and without the enzyme. Both plasma and erythrocytes were used as enzyme sources from both normal subjects and those with a variety of pathologies. The formation of uridine during the reaction was observed.

Erythrocytes from normal subjects were collected, washed, and lysed

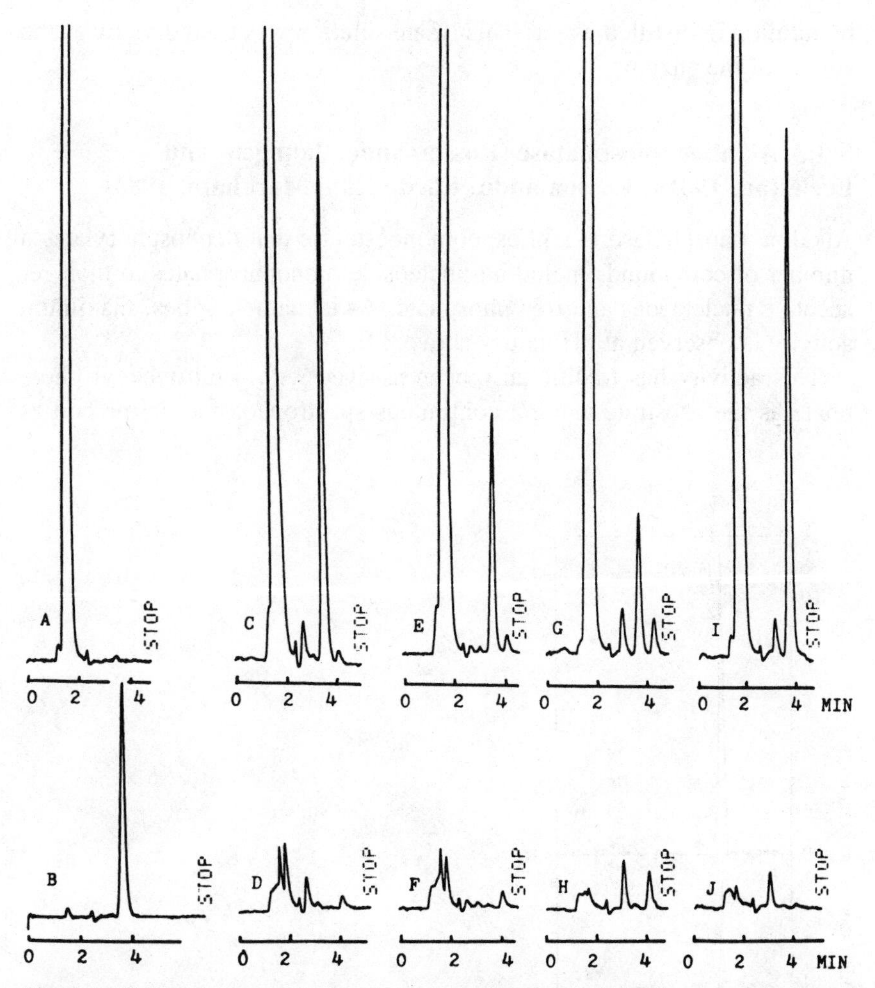

Figure 5.64 Separation of uridine from UMP and blood components. (A, B) Chromatograms of UMP in the reaction mixture without enzyme and uridine standard (50 μM), respectively. Samples were incubated (C, E, G, I) with or (D, F, H, J) without UMP. (C, D) Erythrocytes from a normal subject; (E, F) erythrocytes from a lead-poisoned subject; (G, H) plasma from a normal subject; (I, J) plasma from a person suffering from hepatobiliary disease. (From Sakai et al., 1982.)

by dilution in distilled water. This lysate solution was used directly as the source of the enzyme.

5.9.3 Alkaline Phosphatase (Rossomando, Jahngen, and Eccleston, 1981a; Rossomando, Cordis, and Markham, 1983)

Alkaline phosphatase is a phosphomonoesterase that dephosphorylates a number of compounds including nucleoside monophosphates to their respective nucleosides and free phosphate. As its name implies, maximum activity is observed at pH values above 8.0.

This activity has traditionally been assayed with 4-nitrophenyl phosphate as the substrate using a continuous spectrophotometric method as

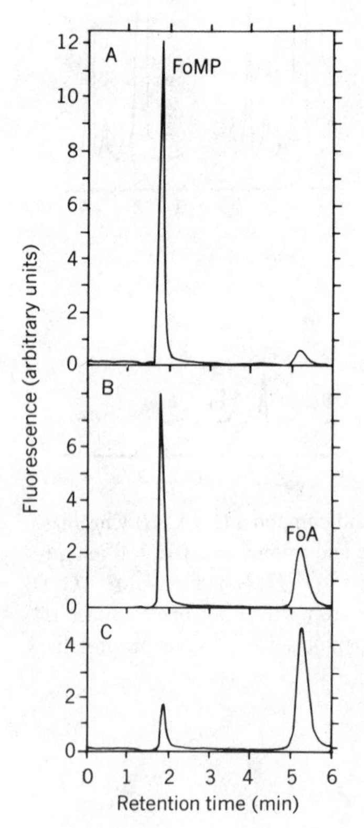

Figure 5.65 The hplc analysis of a reaction mixture containing AMP and alkaline phosphatase. Tracings obtained of reaction mixture (A) immediately after the addition of enzyme, (B) after 10 min, and (C) after 15 min. (From Rossomando et al., 1981a.)

described in Chapter One. Recently, the hplc method has been used together with a nucleoside monophosphate such as AMP as the substrate. The formation of adenosine during the course of the reaction was monitored at 254 nm.

Substrate and product were separated by reverse-phase hplc (μBondapak) using a mobile phase of a phosphate buffer at pH 5.5 with 1% methanol. The column was eluted isocratically, and the detection was at 254 nm.

The reaction mixture contained the substrate and buffer, and the reaction was started by the addition of the enzyme. In one study, the substrate was formycin 5′-monophosphate, a fluorescent analog of AMP (see Chapter Three). The formation of formycin A, the analog of adenosine, is shown in Fig. 5.65 as a function of incubation time.

In an interesting application of this assay to the question of reaction mechanisms, the substrate AMP was replaced by the thioanalog 5′-deoxy-5′-thioadenosine monophosphate [A(S)MP]. The structure of A(S)MP is shown in Fig. 5.66. (This analog is available from Calbiochem-Behring.) With this analog it was possible to explore the question of whether the enzyme cleaved between the C-5′ and the oxygen atoms or between the oxygen atom and the phosphorous atom. These alternatives are illustrated in Fig. 5.67. It is clear that the site of cleavage can be distinguished, since the alternatives will produce different reaction products, in one case thioadenosine and phosphate and in the other thiophosphate and adenosine.

Since thioadenosine is readily separated from adenosine by hplc, the use of this analog together with the hplc assay method allowed the site of cleavage to be established. As shown in Fig. 5.68A, an analysis of the incubation mixture during the course of the reaction revealed the formation of thioadenosine. No adenosine was detected. These findings supported the conclusion that the site of cleavage was the bond between the sulfur and the phosphate.

This analog proved to be useful in another way as well when it was found that the analog was not a substrate for a 5′-nucleotidase, since, as shown in Fig. 5.68B, incubation of the thioanalog with this activity produced no reaction product. These results suggest the possibility that this thioanalog and the hplc assay method may be useful to discriminate between the 5′-nucleotidase and alkaline phosphatase activities.

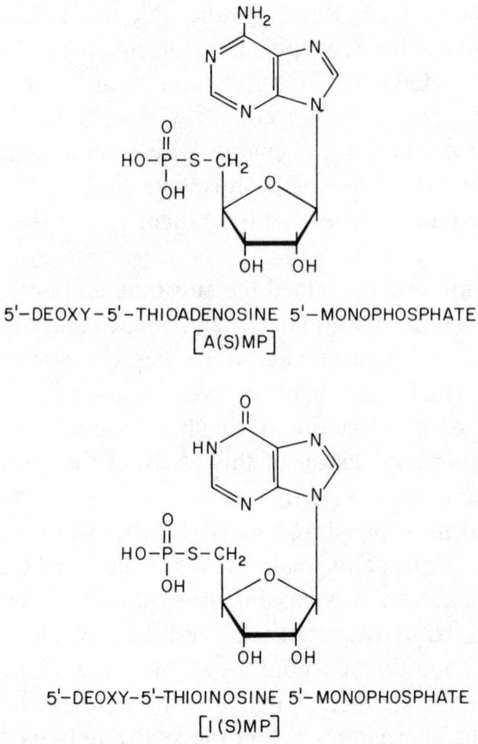

5'-DEOXY-5'-THIOADENOSINE 5'-MONOPHOSPHATE
[A(S)MP]

5'-DEOXY-5'-THIOINOSINE 5'-MONOPHOSPHATE
[I(S)MP]

Figure 5.66 Structure of the thioanalogs of AMP and IMP in which sulfur replaces the bridge oxygen between the 5′ carbon and the phosphorus. (From Rossomando et al., 1983a.)

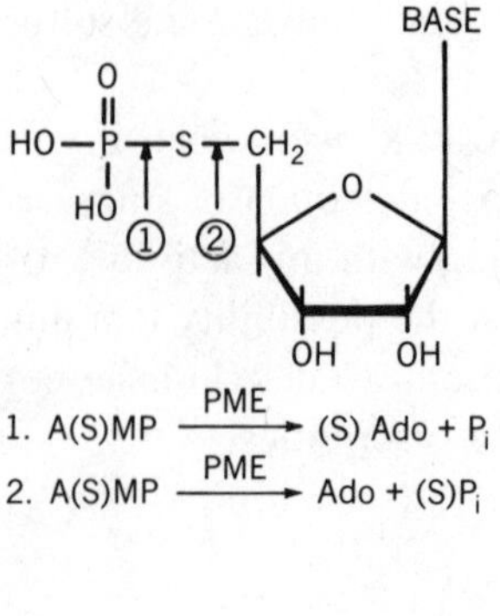

1. A(S)MP $\xrightarrow{\text{PME}}$ (S) Ado + P_i
2. A(S)MP $\xrightarrow{\text{PME}}$ Ado + (S)P_i

Figure 5.67 Sites of bond cleavage by phosphomonoesterases. Arrow 1 indicates cleavage of the bond between the phosphorus and sulfur atoms. The products are shown in reaction (1). Arrow 2 indicates cleavage between the carbon and the sulfur, and the reaction products are shown in reaction (2).

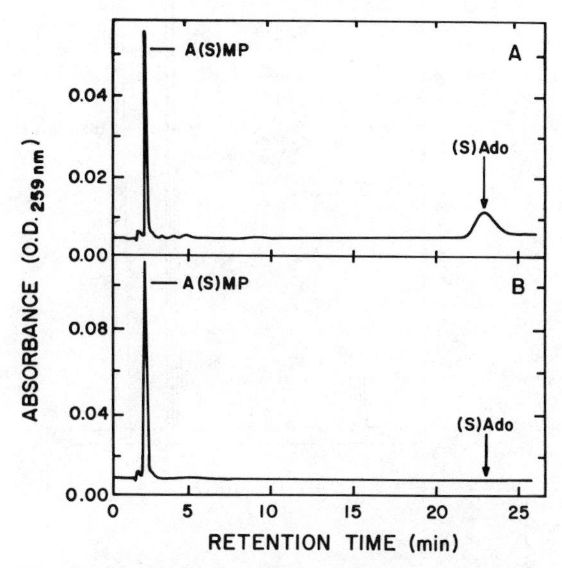

Figure 5.68 Hplc chromatograms of phosphomonoesterase hydrolysis of A(S)MP. (A) Chromatogram obtained from calf intestinal mucosa alkaline phosphatase hydrolysis of A(S)MP. In a reaction volume of 100 μL containing 100 m*M* Tris-HCl (pH 8.1), 300 μ*M* A(S)MP, and 20 m*M* $MgCl_2$, the reaction was initiated by addition of 2 μg of enzyme and incubated at 30°C for 6 h. A 20 μL sample was then injected onto the hplc column and analyzed. (B) Chromatogram obtained from snake venom 5′-nucleotidase incubated with A(S)MP. In a reaction volume of 100 μL containing 100 m*M* Tris-Cl (pH 8.1), 300 μ*M* A(S)MP, and 20 m*M* $MgCl_2$, the reaction was initiated by addition of 6 μg of enzyme and the reaction mixture incubated at 30°C for 60 min, and a 20 μL sample was injected onto the hplc column and analyzed. (From Rossomando et al., 1983a.)

5.9.4 Adenosine Deaminase (Uberti et al., 1977; Hartwick et al., 1978)

Adenosine deaminase catalyzes the deamination of adenosine to form inosine and ammonia. The inosine (Ino) can be degraded further to hypoxanthine (Hyp) by nucleoside phosphorylase, an activity often present in extracts. Therefore, in many cases, the assay involves a determination of either the loss of adenosine (Ado) or the formation of both inosine and hypoxanthine.

In one study the compounds were separated by reverse-phase hplc using columns prepacked with C-18 μBondapak or Partisil ODS. Compounds were eluted isocratically using a mobile phase of methanol and 10

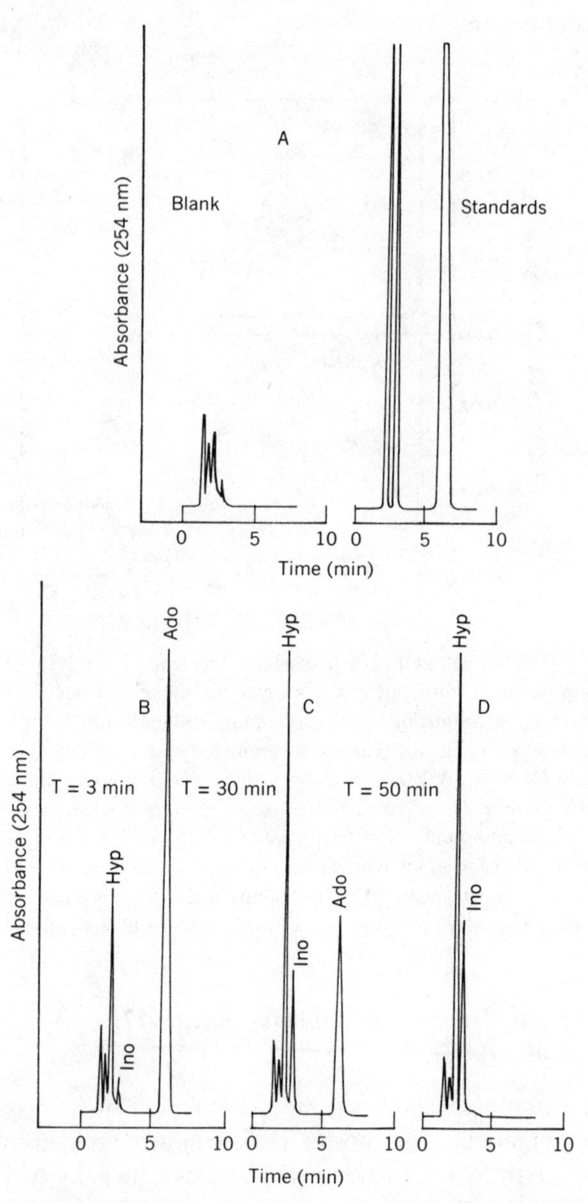

Figure 5.69 The separation of the components of human erythrocytes by reverse-phase hplc. In A, a blank erythrocyte lysate is shown along with three standards: Hyp, Ino, and Ado. In B, C, and D, the decrease in the substrate peak area (Ado) is shown as a function of time. Chromatographic conditions: isocratic elution; flow rate, 2.0 mL/min, mobile phase: 0.01 F KH_2PO_4 (pH unadjusted) and methanol (86:14, v/v). In each chromatogram, the injection volume was 5 μL, at an attenuation of 64 on a Hewlett-Packard integrator. (From Hartwick et al., 1978.)

m*M* KH_2PO_4 (14:86, v/v) with no further pH adjustment. The separations obtained using these conditions are shown in Fig. 5.69A.

The activity was obtained from a lysate of red blood cells. The reactions were terminated with a boiling water bath (45 s) and the samples clarified by centrifugation. Samples of 5 μL were analyzed. In Fig. 5.69B, C, D are shown the chromatographic profiles obtained after incubation times of 3, 30, and 50 min, respectively, with the enzyme. The loss of adenosine is noted, but the effect of the nucleoside phosphorylase is seen, since hypoxanthine and not inosine is the final product.

5.9.5 AMP Deaminase (Jahngen and Rossomando, 1984)

The enzyme adenylic acid deaminase catalyzes the deamination of AMP to IMP and ammonia. For the hplc method, the assay involves the separation of the substrate AMP from the reaction product.

In one study these two were separated by ion-paired reverse-phase hplc on a C-18 (μBondapak) column with a mobile phase of 65 m*M* potassium phospate (pH 3.6), 1 m*M* tetra-*n*-butylammonium phosphate, and 4% methanol. The column was eluted isocratically, and the eluent was monitored at 254 nm. When the formycin analogs were used, detection was at 295 nm; at this wavelength there was no interference from the ATP present in the reaction mixture since the formycin has a maximum at 298 nm and ATP at 265 nm.

The reaction mixture contained, in a final volume of 250 μL, imidazole-HCl (pH 6.8), as the buffer, either 250 nmol AMP or the formycin analog formycin 5′-AMP (FoMP) as substrate, and the activators ATP and KCl. The reaction was started by the addition of the enzyme, and at intervals samples were withdrawn from the reaction tube and injected directly onto the hplc column for analysis.

Figure 5.70 shows a series of chromatograms representing the analysis of samples taken from an incubation mixture. As expected, the chromatogram of the sample of the reaction mixture taken before the start of the reaction shows only the substrate, in this case FoMP. Chromatograms of the incubation mixture sampled after the start of the reaction show the formation of the reaction product formycin 5′-IMP. A comparison of the peak heights of both substrate and product show clearly that the loss of substrate can be completely accounted for by the appearance of product. When these area values were converted to units of concentration, the rate curves shown in the inset to Fig. 5.70 were obtained.

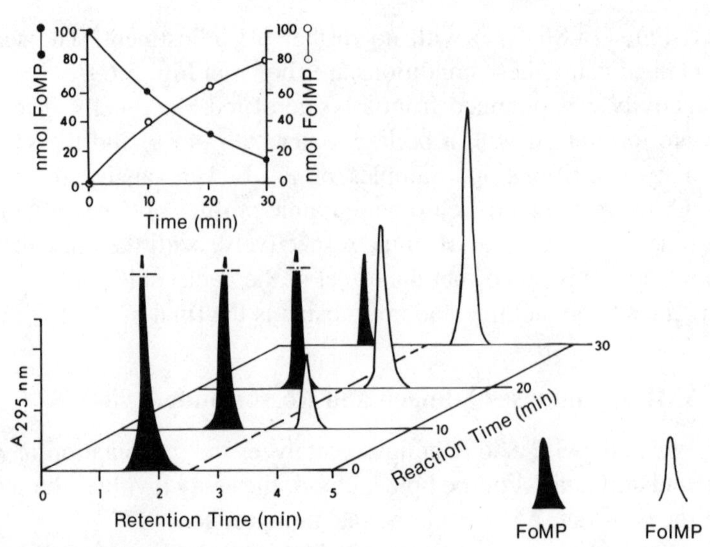

Figure 5.70 Hplc elution profiles of an adenylate deaminase incubation mixture. The reaction mixture contained 15 μmol of imidazole-HCl (pH 6.8), 250 nmol of FoMP, 250 nmol of ATP, and 5 μmol of KCl in a final volume of 250 μL. The reaction was initiated by the addition of activity obtained from the S-100 fraction and incubated at 37°C. At 10 min intervals, 25 μL samples were injected onto the hplc column. There is a decrease in the FoMP peak (retention time, 1.7 min) and a significant rise in the peak corresponding to FoIMP (retention time, 3.1 min) up to 30 min. The inset is a graphical representation of the first 30 min of the reaction. (From Jahngen and Rossomando, 1984.)

The activity was prepared from the microorganism *Dictyostelium discoideum*. Cells were lysed, and an S-100 fraction was prepared (see Chapter Four) and used as the source of the deaminase activity.

5.9.6 Cyclic Nucleotide Phosphodiesterase (Tsukada et al., 1980)

The enzyme 2′,3′-cyclic nucleotide 3′-phosphodiesterase has been suggested as a marker for myelin. The activity catalyzes the degradation of 2′,3′-cyclic AMP to 2′-AMP or 2′,3′-cyclic CMP to 2′-CMP.

These compounds were separated on a Teflon column packed with silica gel-NH_2 (ODS C-18 LiChrosorb NH_2). The mobile phase was a solution of 25 m*M* KH_2PO_4–25 m*M* K_2HPO_4 (1:1, v/v, pH 6.8). Only 1 μL samples were required for analysis. Detection was at 254 nm. The separations obtained with standards on this column are illustrated in Fig. 5.71A.

The reaction mixture contained the diesterase activity from brain homogenate (about 10 μg protein), Tris-HCl buffer (pH 7.4), and $MgSO_4$. The reaction was started by addition of substrate. The reactions

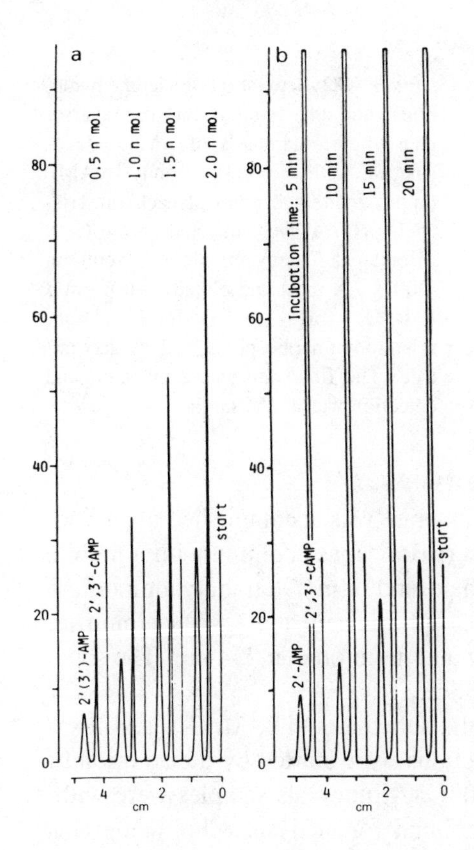

Figure 5.71 Separation and determination of 2′(3′)-AMP. Assay conditions: sample, 1 μL; column, 100 × 0.5 mm, silica gel–NH_2 (ODS-C-18; Merck LiChrosorb NH_2); mobile phase, 8 μL/min; range, 0.16 (absorbance) = 100; chart speed, 1 mm/min. (A) Assay of standard solutions. (B) Assay of enzyme activities. Enzyme: Rat cerebral homogenate (9.5 μg protein). Substrate: 20 m*M* 2′,3′-cAMP. Preincubation: 5 min at 37°C. Incubation: 37°C for 5, 10, 15, 20 min. (From Tsukada et al., 1980.)

were terminated at suitable intervals with ethanol and extracted with chloroform. After centrifugation for 10 min, 1 μL of the supernatant solution was analyzed. In Fig. 5.71B are shown the results of the assay illustrating an increase in the height of the peak of the reaction product, 2′-AMP, with incubation.

Cerebral tissues were homogenized with a glass homogenizer and used as the source of the diesterase activity.

5.9.7 ATP Pyrophosphohydrolase (Rossomando, Jahngen, and Eccleston, 1981a; Rossomando and Jahngen, 1983)

This enzyme catalyzes the hydrolysis of ATP to AMP and pyrophosphate (PP_i). It is an activity that may be involved in several functions including calcification, the polymerzation of actin, or the regulation of ATP levels.

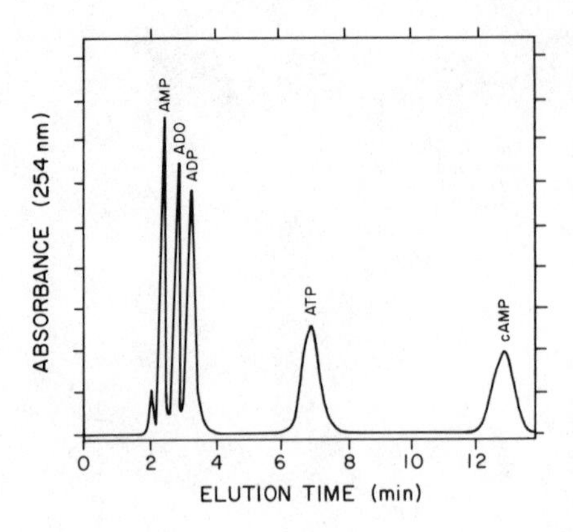

Figure 5.72 Separation of adenine nucleotides and adenosine by ion-pair reverse-phase hplc. Standards of AMP, adenosine, ADP, ATP, and cAMP (approximately 2 nmol of each) in Tris-HCl (pH 7.4) were injected onto a C-18 μBondapak reverse-phase column (300 × 7.8 mm) and eluted with 65 m*M* KH_2PO_4 (pH 3.6), 1 m*M* tetrabutylammonium phosphate, and 2% acetonitrile. The flow rate was 2 mL/min, and detection was at 254 nm.

The separation of product from substrate was accomplished using ion-paired reverse-phase hplc on a C-18 (μBondapak) column with a mobile phase of 65 m*M* potassium phosphate and 1 m*M* tetrabutylammonium phosphate adjusted to pH 3.6 with phosphoric acid and 1.5% acetonitrile. The column was eluted isocratically and monitored at 254 nm. The separations obtained are shown in Fig. 5.72.

The reaction mixture contained the substrate ATP, $MnCl_2$, and a sodium acetate buffer at pH 6.0. Reactions were started by the addition of enzyme, and incubations were at 30°C. At intervals samples were withdrawn and injected onto the hplc column for analysis. Chromatograms obtained showed a time-dependent increase in the amount of AMP and a corresponding decline in the level of the substrate ATP.

The activity was prepared from the microorganism *Dictyostelium discoideum*. Cells were lysed and an S-100 supernatant solution prepared as described in Chapter Four.

5.9.8 Hypoxanthine-Guanine Phosphoribosyltransferase (Ali and Sloan, 1982; Jahngen and Rossomando, 1984)

Hypoxanthine-guanosine phosphoribosyltransferase (HGPRT) catalyzes the formation of IMP and pyrophosphate (PP_i) from hypoxanthine (Hyp) and phosphoribosylpyrophosphate (PRibPP) as shown in reaction (1):

$$\text{Hyp} + \text{PRibPP} \rightarrow \text{IMP} + \text{PP}_i \quad (1)$$

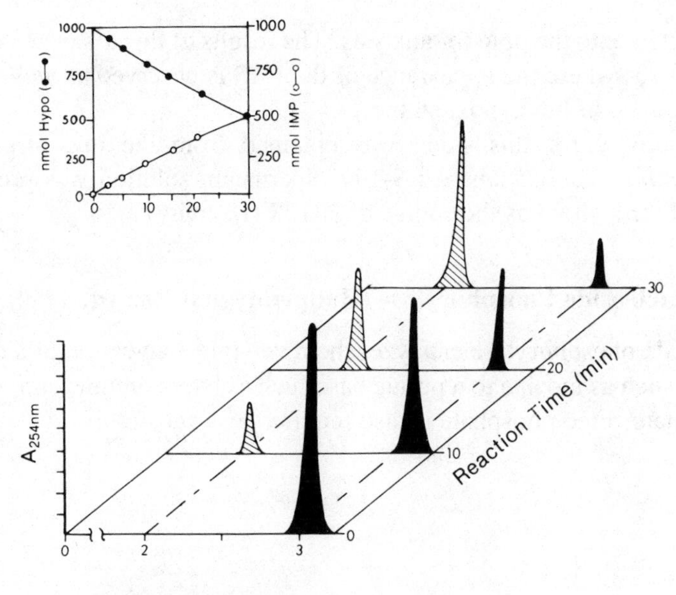

Figure 5.73 Hplc elution profiles of an incubation mixture to study hypoxanthine-guanine phosphoribosyltransferase. The reaction was initiated by the addition of the enzyme mixture, and aliquots were injected onto the hplc column at 10 min intervals as indicated on the *z* axis. The solid peaks represent hypoxanthine, which decreases with time, while the hatched peaks describe the formation of IMP. (From Jahngen and Rossomando, 1984.)

This enzyme represents a principal route for the return, or salvage, of purines such as hypoxanthine, adenine, and guanine to the monophosphate level. The activity requires a metal, preferably magnesium.

Measurement of the activity of this enzyme by the hplc assay method requires separation of the hypoxanthine and the IMP, which can be easily accomplished by several methods including reverse-phase or ion-exchange hplc. In one study, a reverse-phase C-18 column was used with a mobile phase of 10 m*M* potassium phosphate (pH 5.6), and 10% methanol. Detection was at 254 nm, and the separation obtained is shown in Fig. 5.73.

The reaction mixture contained Tris-HCl as buffer, (pH 8.4), $MgCl_2$, and the two substrates hypoxanthine and PRibPP. The reaction was started by the addition of enzyme, and samples were removed at intervals and injected onto the hplc for analysis. The results of this assay are shown in Fig. 5.73, where the appearance of the IMP is observed as well as the disappearance of the hypoxanthine.

The activity for this study was obtained from the microorganism *Dictyostelium discoideum*. An S-100 supernatant solution was prepared and used throughout as the source of HGPRT activity.

5.9.9 Nucleoside Phosphorylase (Halfpenny and Brown, 1980)

Nucleoside phosphorylase catalyzes the reversible conversion of a purine riboside such as inosine to a purine base such as hypoxanthine and ribose-1-phosphate. Free phosphate is also required as a substrate.

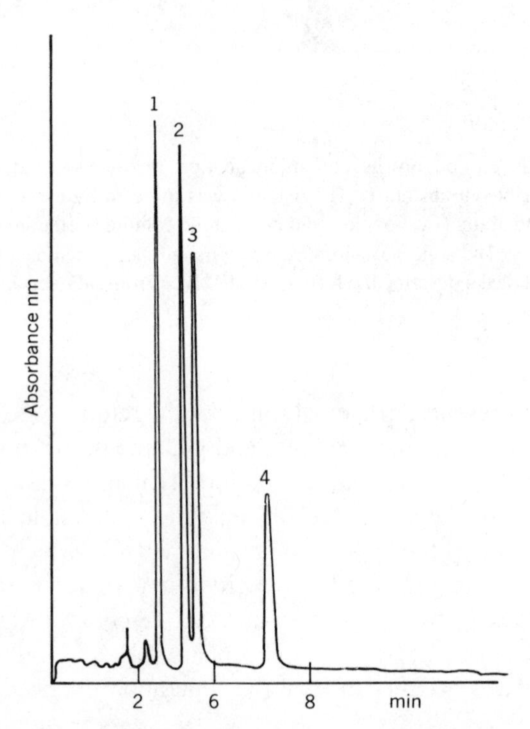

Figure 5.74 Separation of the components of the reaction studied by hplc. Chromatographic conditions: isocratic elution; flow rate, 2 mL/min; 0.02 F KH_2PO_4 (pH 4.2) and 3% methanol. Peaks: (1) uric acid; (2) hypoxanthine; (3) xanthine; (4) inosine. (From Halfpenny and Brown, 1980.)

The assay involves the separation of reactants by reverse-phase hplc on a C-18 (Partisil 5 ODS) column with a mobile phase of 0.02 F KH_2PO_4 (pH 4.2) and 3% methanol applied isocratically. Absorbance measurements were at 254 nm. The separations obtained are shown in Fig. 5.74

The reaction was carried out in the following manner. Stock blood was transferred into test tubes, water was added, and the solution was frozen and thawed to lyse the cells. At zero time, excess xanthine oxidase was added as a coupling enzyme to convert all the hypoxanthine that was formed during the reaction to uric acid. The reaction was started by the addition of inosine in a phosphate buffer.

The complete reaction mixture was incubated for 10 min at 25 °C, and the reaction was terminated by immersing the reaction tubes in a boiling water bath for 1 min. The solutions were clarified by centrifugation, and samples of the supernatant solution were analyzed by hplc.

The reverse reaction was assayed as well, using hypoxanthine and glucose-1-phosphate as substrates in a Tris-HCl buffer at pH 7.4. The results of this assay are shown in Fig. 5.75. The chromatograms, taken at various times, show the decrease in the substrate inosine and the increase in the product uric acid.

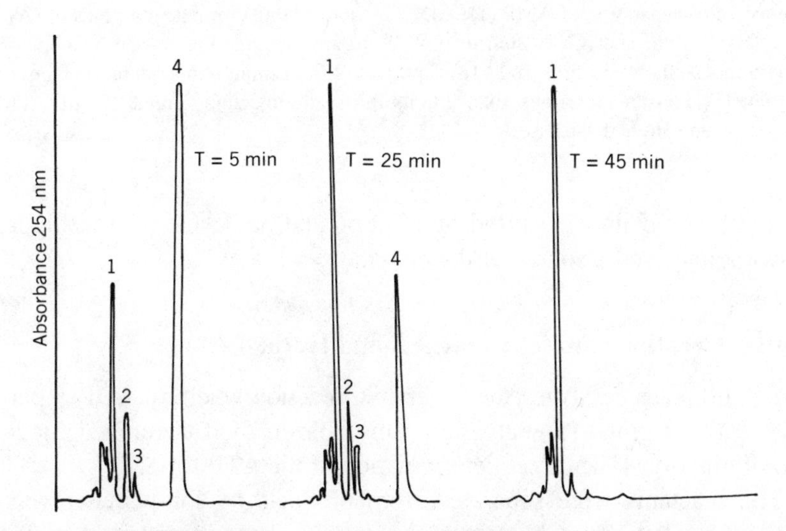

Figure 5.75 Reaction of PNPase as a function of time. Chromatograms at various time intervals show the decrease of the substrate inosine (4) and the increase of the products uric acid (1), hypoxanthine (2), and xanthine (3). (From Halfpenny and Brown, 1980.)

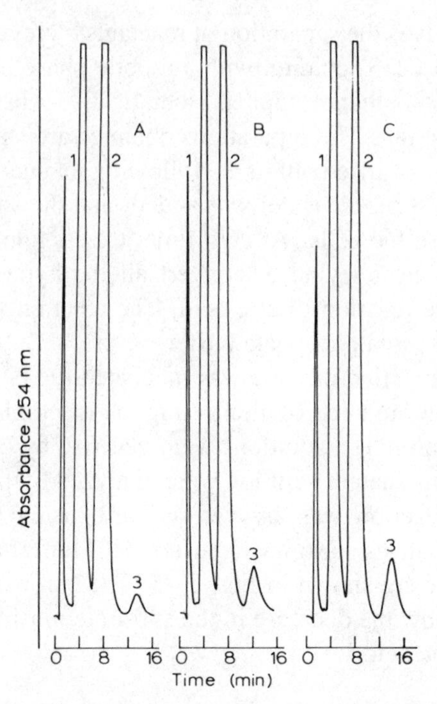

Figure 5.76 Separation of AMP (1), ADP (2), and ATP (3) for reaction times of (A) 10 min, (B) 20 min, and (C) 30 min using 0.04 μg/mL creatine kinase. Chromatographic conditions: column, LiChrosorb C-18; flow rate, 2.0 mL/min; temperature, ambient; detection, UV 254 nm, 0.04 absorbance units full scale; injection volume, 10 μL. (From Danielson and Huth, 1980.)

Red blood cells were lysed by dilution and one cycle of freezing and thawing and used a source of the enzyme.

5.9.10 Creatine Kinase (Danielson and Huth, 1980)

Creatine kinase catalyzes the reversible reaction whereby ADP + phosphocreatine form ATP + creatine. One hplc method developed for this activity involved the direct determination of the ATP formed.

The reactants were separated from products by ion-paired reverse-phase hplc (RP-18 LiChrosorb). The mobile phase consisted of an 88% mixture of 0.1 *M* KH_2PO_4/0.025 *M* butylammonium hydrogen sulfate and 12% methanol. To this was added enough 0.75 *N* NaOH to adjust the pH to 6.8. The separation of ATP, ADP, and AMP is shown in Fig. 5.76.

The reaction mixture contains ADP, AMP, and KF, the last to inhibit any adenylate kinase activity, phosphocreatine and magnesium at concentrations tenfold in excess of ADP. The reaction was started by the addition of enzyme and terminated with a boiling water bath.

After cooling, a 10 μL sample was analyzed by hplc. The results of one assay are shown in Fig. 5.76. The appearance of the ATP is clearly noted and is evidence of creatine kinase activity.

The preparation of creatine kinase was obtained from a commercial source.

5.9.11 Adenosine Kinase (Dye and Rossomando, 1982)

Adenosine kinase catalyzes the transfer of phosphate from ATP to adenosine (Ado) to form AMP and ADP. The separation of the reactants, Ado and ATP, from the products, AMP and ADP, can be accomplished by reverse-phase hplc (C-18) with isocratic elution with a mobile phase of 0.1 *M* potassium phosphate (pH 5.5) and 10% methanol. Detection depends on the substrate. In this assay, it is useful to replace the substrate adenosine with the fluorescent analog formycin A (FoA) and to monitor the column eluent with a fluorescence detector. Thus, ATP and any of its hydrolytic products will not be detected.

The reaction mixture contained Tris-HCl (pH 7.4), as the buffer, ATP, FoA, MgCl, and KCl. The reaction mixture also contained EHNA (*erythro*-9,2-hydroxy-3-nonyl)adenine, an inhibitor of the secondary reaction catalyzed by adenosine deaminase. The reaction was started by the addition of the enzyme preparation and terminated by injecting samples directly onto the hplc column. The results of the reaction are shown in Fig. 5.77.

The enzyme was prepared from mouse liver. The liver was disrupted by grinding, and any insoluble material was removed by centrifugation at 30,000 *g* for 30 min to form an S-30 fraction. This S-30 was used throughout the study.

5.9.12 Adenylate Cyclase (Rossomando, Jahngen, and Eccleston, 1981b)

Adenylate cyclase catalyzes the formation of cAMP and PP_i from ATP. In the hplc assay, cAMP is separated from the ATP substrate by reverse-phase hplc on a C-18 (μBondapak) column with a mobile phase of

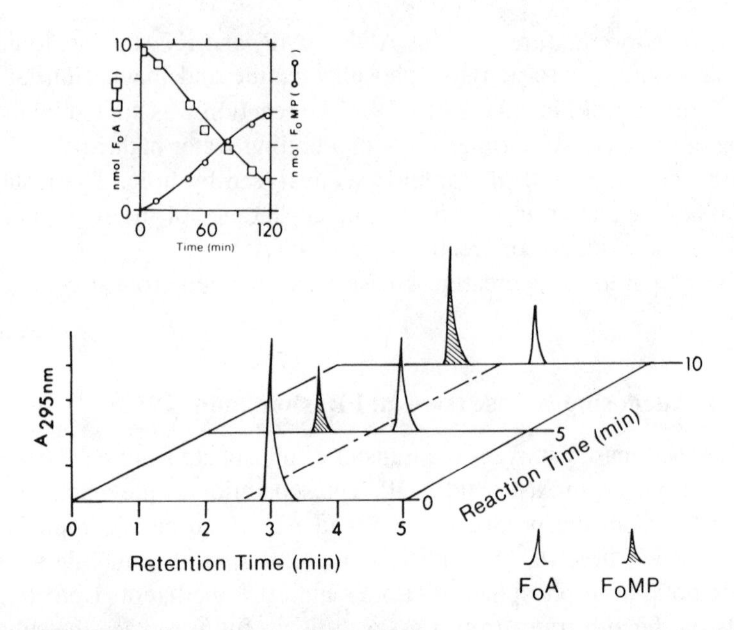

Figure 5.77 Hplc elution profiles of an incubation mixture to study adenosine kinase. The reaction was initiated by the addition of the enzyme mixture (S-100), and samples were removed and analyzed by hplc at 5 min intervals. The inset illustrates the time-dependent utilization of FoA and the formation of FoMP as determined by integration of the respective peaks from the chromatograms. (From Jahngen and Rossomando, 1984.)

0.01 *M* potassium phosphate (pH 5.5) with 10% methanol. The column was eluted isocratically. The detectors for the eluent depended on the substrate (see below).

The separations obtained are shown in Fig. 5.78 for the fluorescent analog of ATP, formycin ATP (FoTP), which was used instead of ATP because of the greater sensitivity of fluorescence detectors. Therefore, by substituting a fluorometer for the UV spectrophotometer, an increased sensitivity of five- to tenfold was achieved and lower levels of the reaction product could be detected. A calibration curve obtained with formycin is shown in Fig. 5.79.

The reaction mixture contained FoTP and $MgCl_2$, with Tris-HCl (pH 7.5) as the buffer. The reaction was started by the addition of the enzyme, and samples were removed at intervals and injected onto the hplc column for analysis. Chromatograms were obtained, and the cFoMP peak was in-

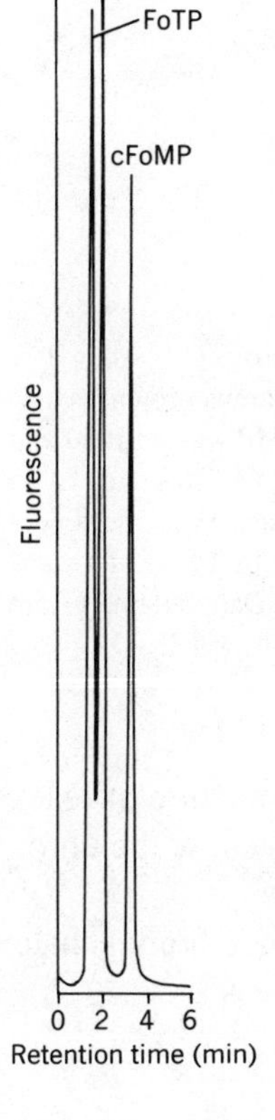

Figure 5.78 Separation of FoTP, FoMP, and cFoMP by hplc. Operating conditions: 0.01 *M* KH_2PO_4 (pH 5.5) buffer with 10% methanol; flow rate, 2 mL/min; room temperature; μBondapak C-18 column packing; fluorescence, excitation at 300 nm and emission above 320 nm. Approximately 10 μg of each component was injected. (From Rossomando et al., 1981a.)

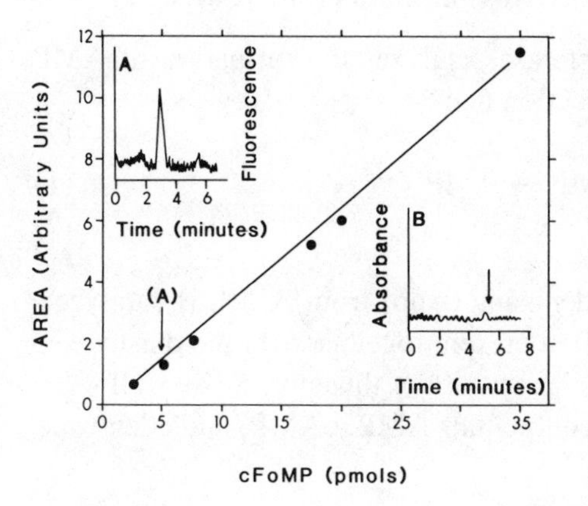

Figure 5.79 Calibration of fluorometric detector with cFoMP. Solutions were prepared ranging from 0.4 to 7 μ*M*, and 5 μL of each was injected onto the column. Areas of the cFoMP peaks were determined from tracings obtained at each concentration and are plotted (in arbitrary units) as a function of amount of cFoMP injected. (Inset A) Tracing obtained after injection of 5 pmol of cFoMP. (Inset B) Tracing obtained with 15 pmol of cAMP (arrow indicates retention time of authentic cAMP). (From Rossomando et al., 1981b.)

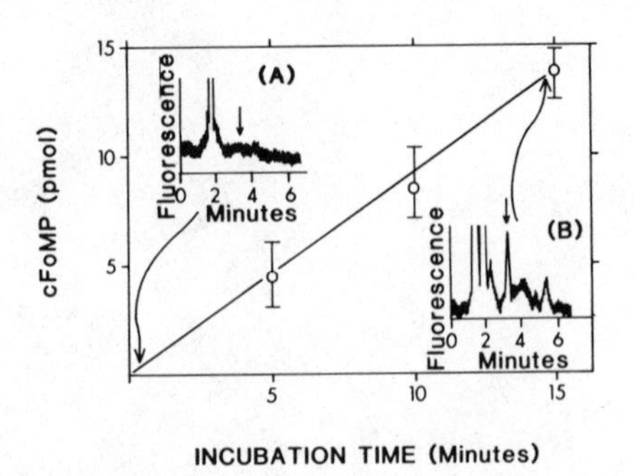

Figure 5.80 Kinetics of cFoMP formation as determined by fluorometry. Adenylate cyclase activity was determined at 0.3 m*M* FoTP with 100 μg of membrane protein in a final reaction volume of 100 μL. Reactions were terminated, and cFoMP was purified and then analyzed by hplc. Insets: Representative hplc profiles obtained at (A) 30 s and (B) 15 min after start of the reaction. The arrows indicate the retention time observed after injection of authentic cFoMP. The area under the curves was determined by integration, and amount of cFoMP present was determined from a calibration curve. Data obtained from hplc assay gave values within the error bars. (From Rossomando et al., 1981b.)

tegrated to determine the amount of cFoMP formed as a function of reaction time. These data are shown in Fig. 5.80. The insets show the chromatograms obtained on two samples.

The membrane-bound activity was obtained from a crude homogenate of rat osteosarcoma cells.

5.9.13 cAMP Phosphodiesterase (Rossomando et al., 1981a)

The enzyme cAMP phosphodiesterase catalyzes the formation of AMP from cyclic AMP as shown in reaction (1):

$$\text{cAMP} \rightarrow \text{AMP} \tag{1}$$
$$\text{AMP} \rightarrow \text{Ado} + \text{P}_i \tag{2}$$

Reaction (2), the formation of adenosine (Ado) from AMP, is catalyzed by 5′-nucleotidase, an activity often present together with the diesterase. It is useful to be able to measure the activity of this enzyme as well.

In the hplc assay method, the compounds AMP, cAMP, and adenosine

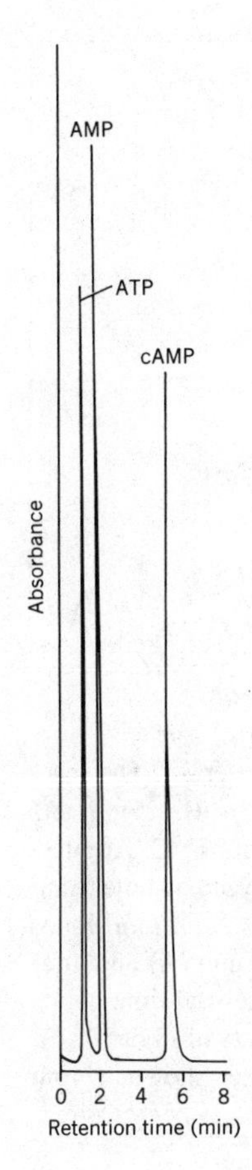

Figure 5.81 Separation of ATP, AMP, and cAMP by hplc. Operating conditions: 0.01 *M* KH_2PO_4 (pH 5.5) buffer with 10% methanol; flow rate, 2 mL/min; room temperature, μBondapak C-18 column packing; absorbance at 260 nm. Approximately 10 μg of each component was injected. Sensitivity of detection, 0.5 absorbance units full scale. (From Rossomando et al., 1981a.)

are separated by reverse-phase hplc on C-18 (μBondapak) with a mobile phase of 10 m*M* KH_2PO_4 (pH 5.5) containing 1% methanol. The separation of cAMP and AMP obtained with a mobile phase of a phosphate buffer and 10% methanol is shown in Fig. 5.81. Adenosine elutes at about 10 min (not shown), which clearly allows the measurement of the level of each of these components in an incubation mixture during the course of reaction.

The reaction mixture contained cyclic formycin monophosphate, an analog of cAMP, as the substrate, Tris-HCl (pH 7.5) as buffer, and $MgCl_2$. The reaction was started by the addition of the enzyme. Samples

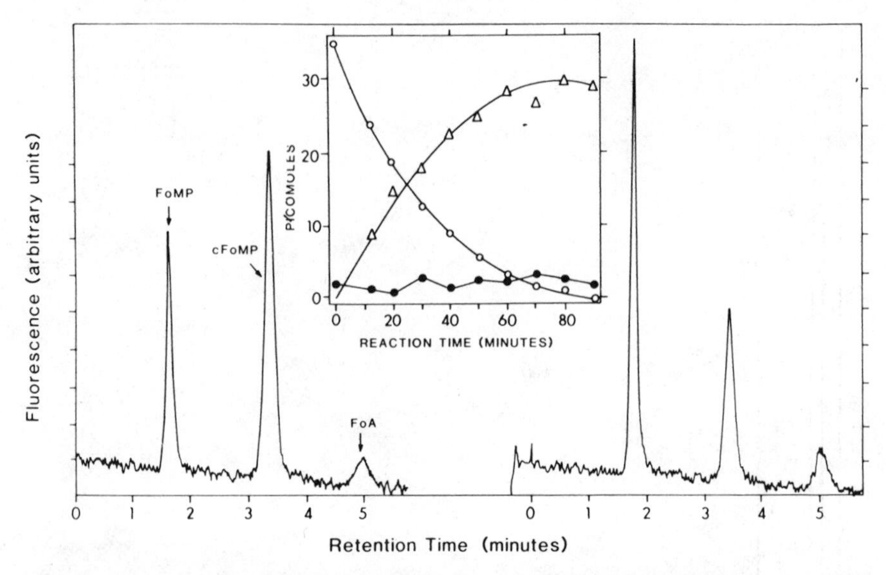

Figure 5.82 Degradation of cFoMP by 3′,5′-cyclic AMP phosphodiesterase. The reaction mixture contained, in a final volume of 50 μL, 10 m*M* Tris-HCl (pH 7.5), 1 μ*M* $MgCl_2$, 1 m*M* cyclic nucleotide. and 50 μg enzyme. Incubations were at 37° C. At intervals, samples were removed from the reaction mixture (20 μL) and analyzed by hplc using a μBondapak C-18 column preceded by a guard column. Flow rate was 2 mL/min. Fluorescence of column effluent was monitored. Profiles obtained after 10 min (A) and after 30 min (B) incubation. Areas under cFoMP peak were determined at several time points from profiles such as those shown in (A) and (B). Inset shows amounts of cFoMP (○), FoMP (Δ), and FoA (●) as determined from tracings obtained at times shown. (From Rossomando et al., 1981a.)

were removed at intervals and injected directly onto the reverse-phase column for analysis. Figure 5.82 shows two chromatograms after 10 and 30 min of incubation. A comparison of these two chromatograms shows that while the amount of cFoMP substrate in the incubation mixture has declined and the amount of product FoMP has increased, the amount of formycin A (FoA), the analog of adenosine, has remained unchanged. When the area of each peak is plotted as a function of reaction time, the data shown in the central inset are obtained. While these data clearly illustrate the activity of the cyclic phosphodiesterase, they also show the absence of any 5′-nucleotidase.

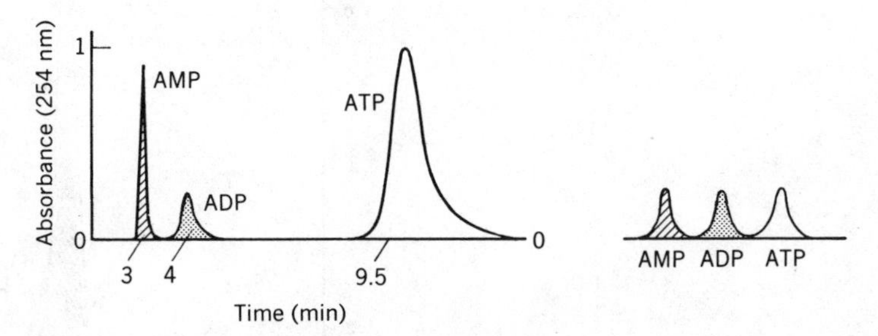

Figure 5.83 Separation of adenine nucleotides and adenosine by paired-ion reverse-phase hplc. Standards of AMP, adenosine, ADP, ATP, and cAMP (approximately 2 nmol each) in Tris-HCl (pH 7.4) were injected onto a C-18 μBondapak reverse-phase column (300 × 7.8 mm) and eluted with 65 m*M* KH_2PO_4 (pH 3.6), 1 m*M* tetrabutylammonium phosphate, and 3% acetonitrile. The flow rate was 2 mL/min, and detection was at 254 nm. (From Rossomando and Jahngen, 1983.)

The activity was purified from *Dictyostelium discoideum* by differential centrifugation and gel filtration.

5.9.14 Adenylate Kinase (Rossomando and Jahngen, 1983)

Adenylate kinase (myokinase) catalyzes the reversible reaction shown in (1):

$$\text{ATP} + \text{AMP} \rightarrow 2\text{ADP} \tag{1}$$

In one assay developed to measure this activity, ion-paired reverse-phase hplc (C-18) was used to separate the reactants from the products. The mobile phase was 0.065 *M* potassium phosphate with 1 m*M* tetrabutylammonium phosphate adjusted to pH 3.6 with phosphoric acid and 3% acetonitrile. The column was eluted isocratically and monitored at 254 nm and with a radiochemical detector. The reaction mixture contained Tris-HCl buffer, ATP, and AMP, and, when assayed in the reverse direction, ADP was the substrate. Under these conditions, the separation shown in Fig. 5.83 was obtained.

The reaction mixtured contained the substrates unlabeled ATP and radioactive [^{3}H]AMP, buffered with Tris-HCl (pH 7.4). The reaction was

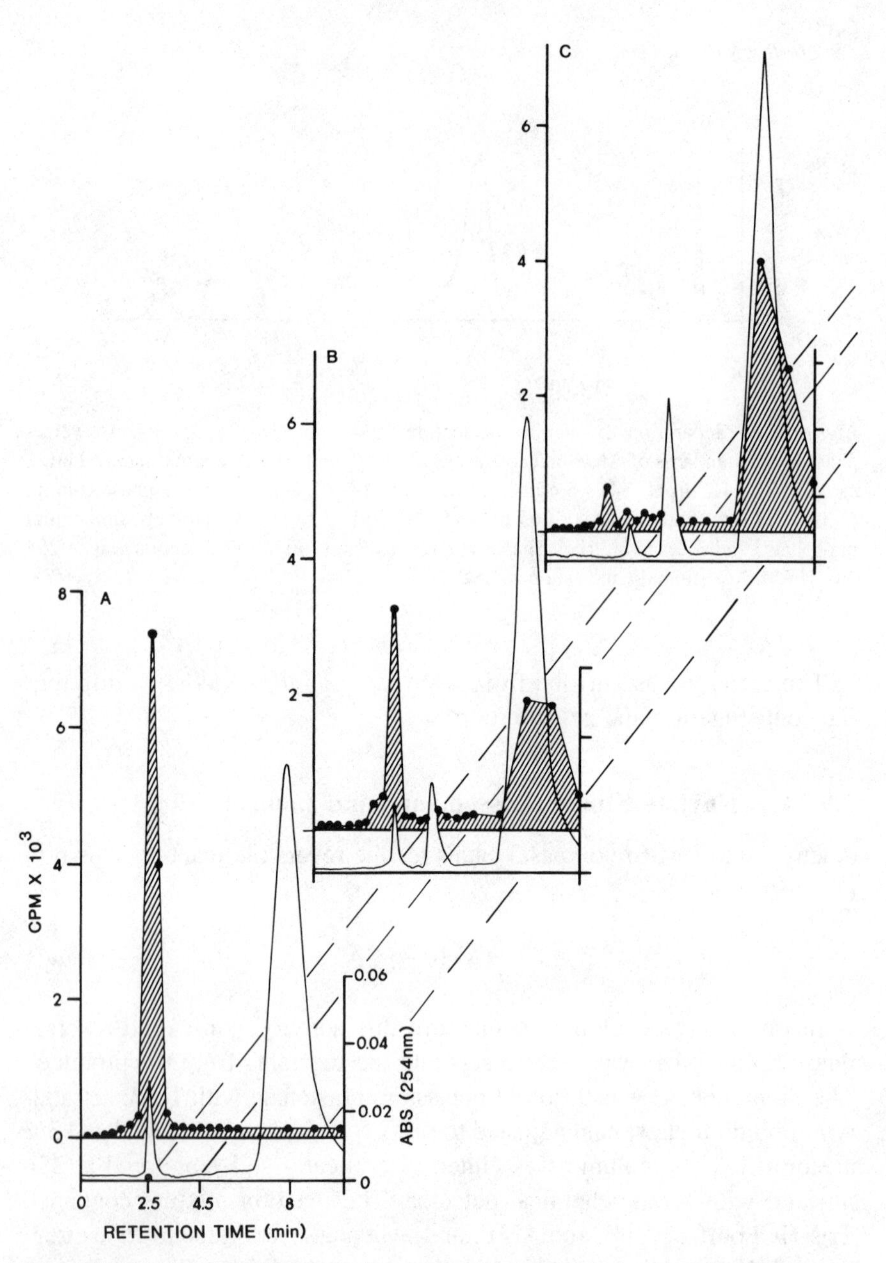

Figure 5.84 Adenylate kinase activity measured by the hplc assay method. The assay mixture contained 200 μ*M* ATP, 20 μ*M* [^{3}H]AMP (approximately 120,000 cpm), 50 m*M* Tris-HCl (pH 7.4), and 32 μg of protein of the enzyme from *D. discoideum*. Samples (20 μL) were injected and the radioactivity monitored continuously with a Berthold LB 503 detector using PICO-Fluor 30 scintillant. Absorbance was measured at 254 nm. Three representative time points were shown, (A), 1 min, (B), 30 min and (C) 60 min after intiation of the reaction.

started by addition of enzyme and terminated by injection of samples directly onto the hplc column.

The results are summarized in Fig. 5.84. Panel A shows the chromatographic profile of the reaction mixture before the addition of the enzyme: the two substrates in the reaction mix, [^{3}H]AMP and ATP, are detected. After initiating the reaction, ADP is formed, and after the reaction proceeds in the reverse direction the amount of radioactive ATP increases steadily until almost all the radioactivity originally present in the AMP is recovered in the ATP.

The activity was obtained from *Dictyostelium discoideum*.

5.9.15 Adenylosuccinate Synthetase (Jahngen and Rossomando, 1984)

The enzyme adenylosuccinate synthetase condenses IMP with aspartic acid to form adenylosuccinate (sAMP). GTP participates directly in the reaction process, and during the course of the reaction GDP is formed.

For the hplc assay, IMP, sAMP, GTP, GDP, and AMP are separated by ion-paired reverse-phase hplc (C-18) with a mobile phase of 65 m*M* potassium phosphate (pH 4.4), 1 m*M* tetrabutylammonium sulfate, and 10% methanol. Detection was at 254 nm. A representative chromatogram is shown in Fig. 5.85.

In addition to the enzyme the reaction mixture contained HEPES buffer, IMP, GTP, MgCl, and creatine phosphate and phosphocreatine kinase (a regeneration system for GTP). The reaction was initiated by adding aspartic acid. Samples were removed at intervals, and the reactions were terminated by direct injection onto the hplc column. In Fig. 5.86 are

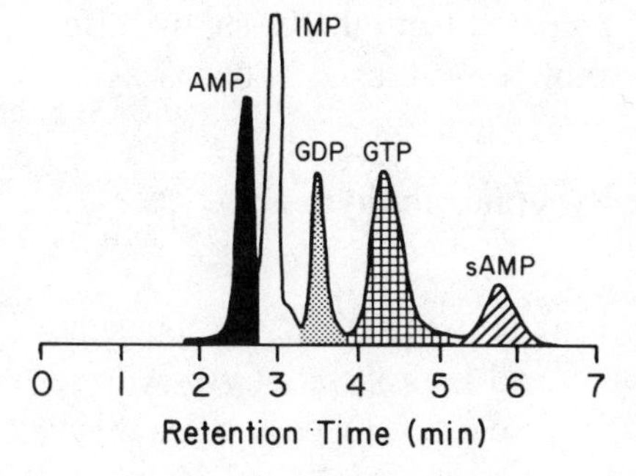

Figure 5.85 Separation of substrates and products of reaction catalyzed by adenylosuccinate synthetase. Column: Prepacked C-18 μBondapak, 10 μm particle size. Mobile phase: 65 m*M* potassium phosphate, 1 m*M* tetrabutylammonium phosphate, 10% methanol at pH 4.4. Absorbance was measured at 254 nm.

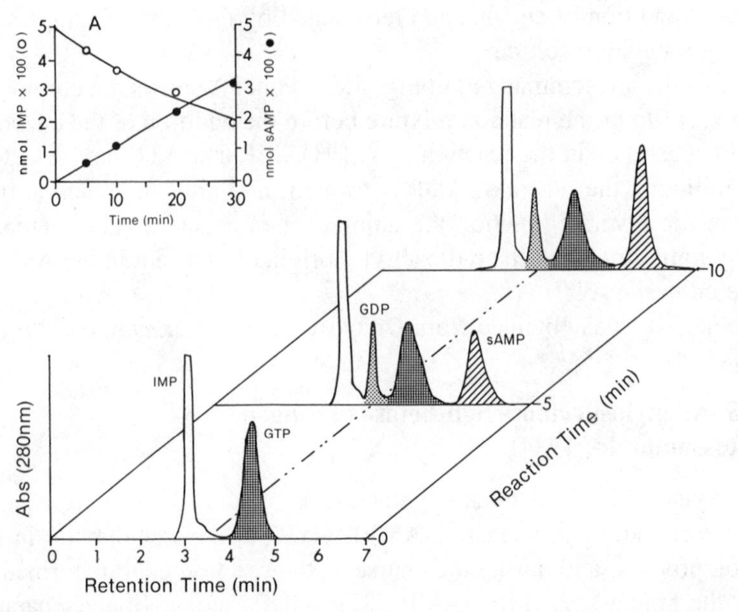

Figure 5.86 Hplc elution profiles of adenylosuccinate synthetase incubation mixtures. This reaction was initiated by the addition of 1.25 μmol of aspartate (pH 7.4). At 5 min intervals, 20 μL samples were injected onto the hplc reverse-phase column and eluted. The inset illustrates the time-dependent utilization of IMP and the formation of sAMP, as determined by integration of the respective peaks from the hplc chromatograms. (From Jahngen and Rossomando, 1984.)

shown chromatograms of samples removed at 0, 5, and 10 min of incubation. The disappearance of IMP and GTP and the appearance of GTP and sAMP are shown.

The enzyme was prepared from *Dictyostelium discoideum*. The cells were lysed, and S-100 solutions were prepared from the lysate by differential centrifugation. Samples of this solution were used in the assay.

5.9.16 Dinucleoside Polyphosphate Pyrophosphohydrolase (Garrison et al., 1982)

This activity (and the specific diadenosine tetraphosphate pyrophosphohydrolase activity) catalyzes the symmetrical hydrolysis of Ap_4A to release 2 molecules of ADP. In the hplc method developed for this assay,

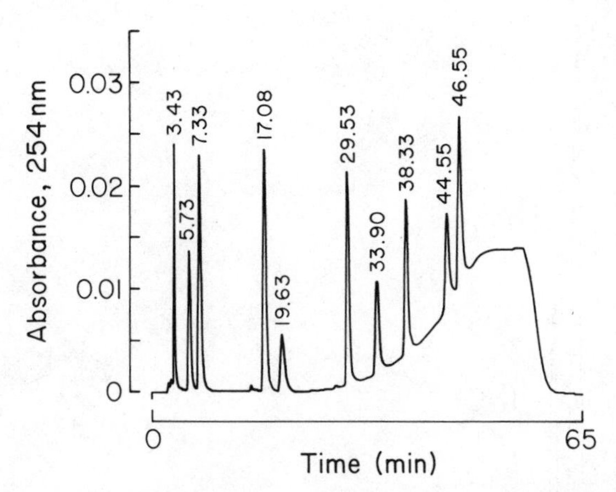

Figure 5.87 Chromatographic resolution of diadenine and monoadenine nucleotides by hplc. 0.8–1.2 nmol of each were separated on a 4.6 mm × 25 cm column of Whatman Partisil PXS 10/25 SAX resin with buffer A, 50 m*M* ammonium phosphate (pH 5.2) and buffer B, 1 *M* ammonium phosphate (pH 5.7) by gradient elution. Flow rate, 1.5 mL/min; temperature, 23–24°C. Nucleotides and retention times: cAMP, 3.43 min; AMP, 5.73 min; Ap_2A, 7.33 min; Ap_3A, 17.08 min; ADP, 19.63 min; Ap_4A, 19.53 min; ATP, 33.9 min., Ap_5A, 38.33 min; Ap_4, 44.55 min; Ap_6A, 46.55 min. (From Garrison et al., 1982.) [Reprinted with permission from Biochemistry, 21: 6129–6133 (1982) American Chemical Society.]

the substrate was separated from the product on an anion-exchange column (Partisil PXS 10/25 SAX), and the amounts of both compounds determined. A precolumn (Solvecon) was placed in line between the mixing chamber and injection valve for mobile phase conditioning. Nucleotides were eluted with a pH and ionic gradient as follows: The column was initially equilibrated with 50 m*M* ammonium phosphate (pH 5.2). After sample injection, the concentration of the second buffer, composed of 1 *M* ammonium phosphate (pH 5.7), was brought to 5%. After 10 min, the concentration of the second buffer was increased linearly from 5 to 45% over a 30-min period. The separation obtained is shown in Fig. 5.87.

The reaction mixture contained HEPES-NaOH (pH 7.5), the substrate (Ap_4A), and the enzyme. After a 10 min incubation, the reactions were stopped by quick-freezing on dry ice or by injecting samples directly onto the hplc column.

The enzyme was purified to homogeneity from *Physarum polycephalum*.

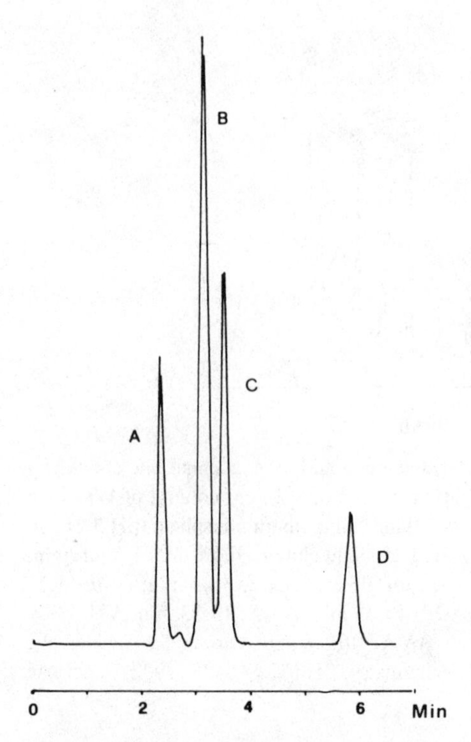

Figure 5.88 Hplc chromatogram of ADPR (A), NAD^+ (B), 4-aminobenzoic acid (C), and nicotinamide (D). (From Pietta et al., 1983.)

5.9.17 NAD Glycohydrolase (Pietta et al., 1983)

The NAD glycohydrolase (EC 3.2.2.5) in this study catalyzes the hydrolysis of NAD^+ to form nicotinamide, adenosine diphosphate ribose (ADPR), and H^+. The assay developed for this activity follows the disappearance of the substrate NAD^+ and the production of nicotinamide.

The separations were accomplished by reverse-phase hplc on a C-18 (μBondapak) column with a precolumn packed with C-18 Corasil, using a mobile phase of 0.01 *M* diammonium hydrogen phosphate–acetonitrile (100:5) at pH 5.5 adjusted with 20% phosphoric acid. Detection was at 259 nm. The separation obtained is shown in Fig. 5.88.

The reaction mixture contained NAD in 0.2 *M* phosphate buffer (pH 7.0), and was equilibrated at 37°C. The reaction was started by addition of the hydrolase, and at intervals it was terminated by diluting a sample in the mobile phase. Each sample was then analyzed by hplc. Intervals as short as 6 min (the time required for a complete hplc run and quantitation) were used. For shorter intervals the reaction was terminated with 1 *M* HCl (pH 3.0) and diluted with mobile phase prior to injection.

The hydrolase was prepared from *Neurospora crassa*. A crude extract was obtained, solubilized by extraction with KCl, and used throughout the study.

5.9.18 Assay of Enzymes Involved in Cytokinin Metabolism (Chism et al., 1984)

Cytokinins, N^6-substituted derivatives of adenine, are important in the regulation of many processes in plant tissues. In a recent paper, an hplc assay was reported measuring the hydroxylation of 6-(3-methylbut-2-enylamino)purine (IPA) and 6-(3-methylbut-2-enylamino)-8-hydroxypurine by cytokinin nucleosidases and adenosine nucleosidases.

The separations were carried out by reverse-phase hplc (C-18 μBondapak) using a mobile phase of 40% methanol–water. The column was eluted isocratically and monitored at 275 nm.

Reaction conditions for hydroxylation of IPA included a phosphate buffer (pH 7.4), and the partially purified xanthine oxidase added to start the reaction. The reaction was stopped at 10 min intervals with methanol, clarified by centrifugation, filtered, and analyzed by hplc. The results of an assay are shown in Fig. 5.89.

The xanthine oxidase for the hydroxylation was obtained by previously published procedures.

5.10 OXYGENATIONS

5.10.1 Acetanilide 4-Hydroxylase (Guenthner et al., 1979)

Acetanilide 4-hydroxylase (A4H) is a microsomal monooxygenase activity that can be followed using the model substrate acetanilide. In the hplc assay developed for this activity, the conversion of acetanilide to 4-hydroxyacetanilide was followed.

The separation of the substrate and the product was accomplished by reverse-phase hplc on an ODS column. The column was eluted isocrati-

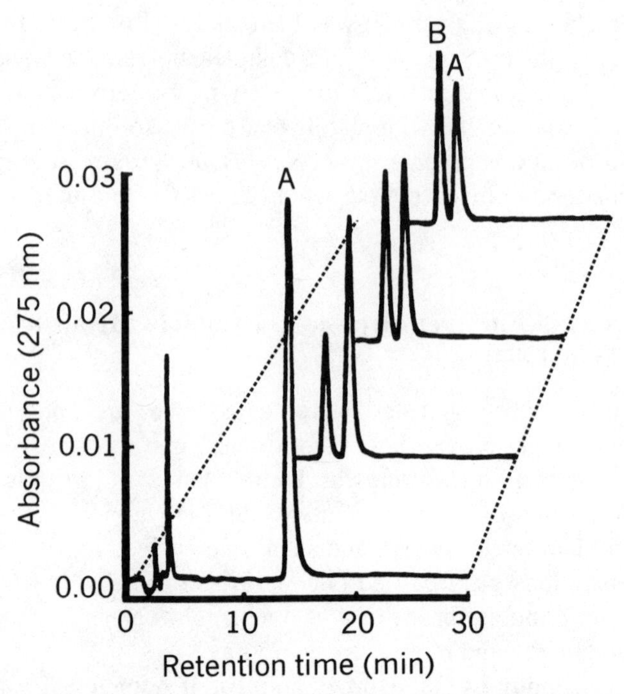

Figure 5.89 Reverse-phase hplc for monitoring the formation of 8-OH IPA from IPA. Aliquots (20 μL) of the stopped reaction mixture were injected onto a μBondapak C-18 column and eluted with 40% aqueous methanol at a flow rate of 1 mL/min. IPA (peak A) is eluted after 8-OH IPA (peak B), which is absent at zero time. The chromatograms are from aliquots taken at 0, 30, 60, and 90 min. (From Chism et al., 1984.)

cally with a mobile phase of methanol–water (33:67 v/v). The compounds were detected at 254 nm. The separations obtained are shown in Fig. 5.90. Radiolabeled substrates were also used, and the eluent was assayed for radioactivity on fractions collected during the elution.

The reaction mixture contained Tris-HCl buffer (pH 7.6), $MgCl_2$, water, NADPH, and NADH. The microsomal suspension was added, followed by radiolabeled acetanilide to start the reaction. After 10 min, the reaction was terminated with ice-cold ethyl acetate, extracted, dried, and analyzed by hplc. A chromatogram showing the formation of product is shown in Fig. 5.91, and the formation of product with time of incubation in Fig. 5.92.

The enzyme activity was obtained using a microsomal suspension prepared from mouse liver.

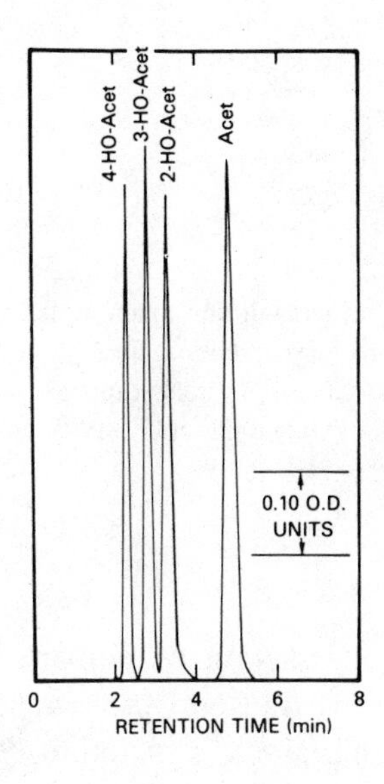

Figure 5.90 Hplc chromatogram of 4-hydroxy-, 3-hydroxy-, and 2-hydroxyacetanilide and acetanilide. About 12.5 μg of each of the four compounds in a total volume of 25 μl of methanol was injected. Approximately 0.33 mL fractions were collected. (From Guenthner et al., 1979.)

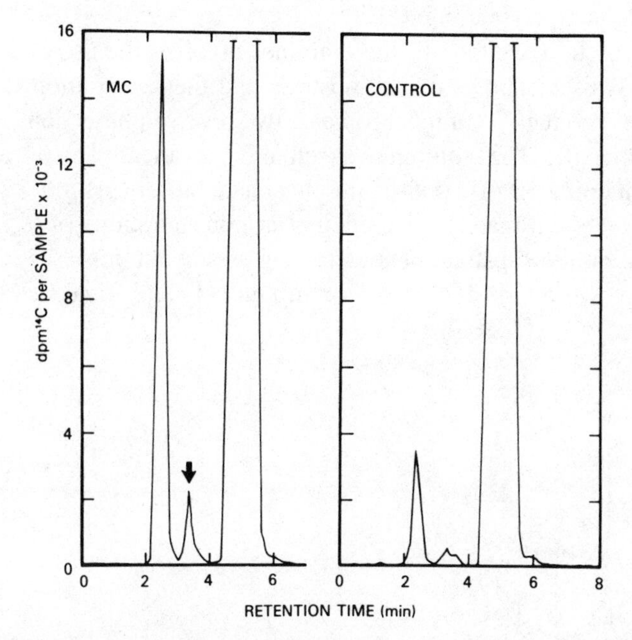

Figure 5.91 Hplc chromatogram of acetanilide and its metabolites generated by hepatic microsomes from mice induced with 3-methylcholanthrene (left) and control (right) mice. Approximately 0.33 mL fractions were collected. Arrow indicates formation of 2-OH acetanilide. (From Guenthner et al., 1979.)

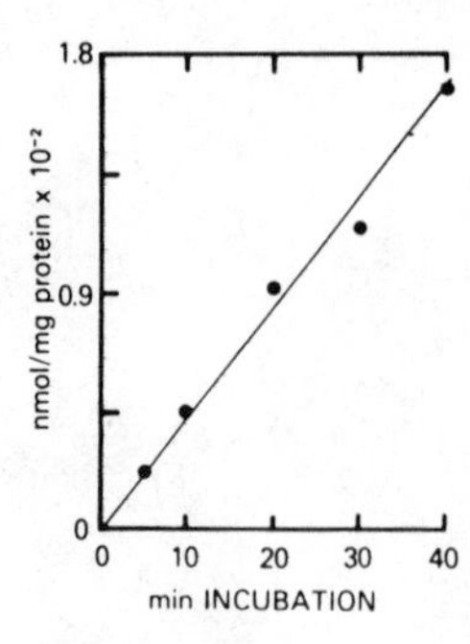

Figure 5.92 Formation of 4-hydroxyacetanilide in nanomoles per milligram protein as a function of incubation time. For these studies, the entire 4-hydroxyacetanilide peak (approximately 1.5 mL) was collected. (From Guenthner et al., 1979.)

5.10.2 Ceruloplasmin (Richards, 1983)

Ceruloplasmin, a protein from the alpha$_2$-globulin fraction of human plasma, is usually considered to be the major copper-transport protein. However, it also catalyzes the oxidation of biogenic amines, including catecholamines, adrenaline, noradrenaline, dopamine, and the indoleamine 5-hydroxytryptamine (5HT).

The assay for oxidation of these amines involves the use of any of the compounds listed above as the substrate and their separation from their respective products, aminochromes, by reverse-phase ion-pair hplc (ODS-Hypersil). The column was eluted isocratically using a mobile phase containing 50 m*M* potassium phosphate buffer (pH 5.5), 2 m*M* sodium heptane sulfonate, and methanol at concentrations of 7.5–17.5% depending on the substrate. The detection was at 300 nm.

Activity was assayed in a reaction mixture (1.2 mL final volume) con-

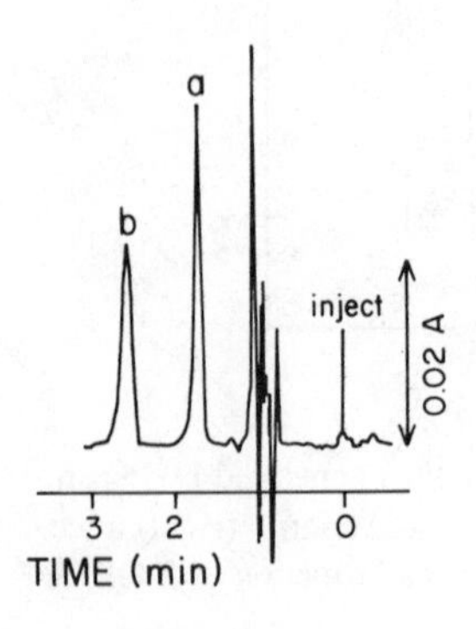

Figure 5.93 Chromatogram obtained following incubation of adrenaline with ceruloplasmin. Peaks: (a) adrenochrome; (b) adrenaline. (From Richards, 1983.)

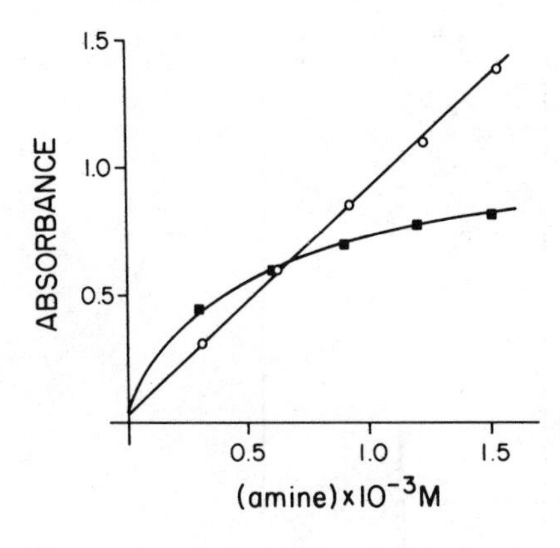

Figure 5.94 Relationship between amine concentration and oxidation product formation at a fixed level of ceruloplasmin. Oxidation of adrenaline (■) was monitored at 300 nm; oxidation of 5-hydroxytryptamine (○) was monitored at 315 nm. (From Richards, 1983.)

taining the substrate and buffer. The reaction was started by the addition of the enzyme and, after incubation at 37°C for 45 min, was terminated by the addition of a 15 m*M* sodium azide solution. Samples were removed and injected directly onto the hplc column for analysis. To generate products to be used as standards, oxidation of each of the amines to the corresponding aminochrome was carried out by incubation of the substrate with 30 m*M* potassium hexacyanoferrate(III) solution for 1 h at room temperature.

In Fig. 5.93, a chromatogram of a sample following incubation of adrenaline with ceruloplasmin is shown. The appearance of the peak of the reaction product, adrenochrome, indicates the activity of the enzyme. Figure 5.94 shows product formation following the oxidation of adrenaline and 5HT.

The enzyme was obtained from commercial sources or from human serum.

5.10.3 Aryl Hydrocarbon Hydroxylase (EC 1.14.14.2) (Tulliez and Durand, 1981)

Aryl hydrocarbon hydroxylase (AHH) is part of the microsomal mixed function oxidase system involved in the detoxification of polycyclic aromatic hydrocarbons. In the hplc assay developed for the AHH activity, benzo[*a*]pyrene (B*a*P) is used as the substrate, and the activity is determined by measuring the unreacted B*a*P during the reaction.

The quantitation of B*a*P was accomplished by reverse-phase hplc (Li-

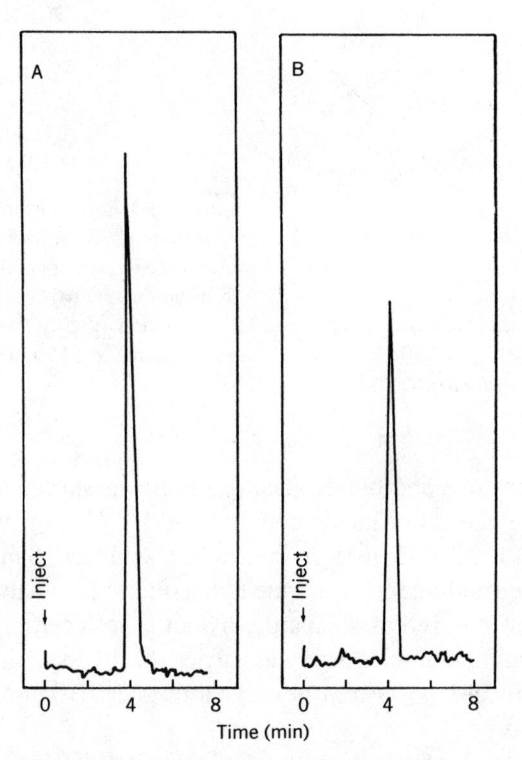

Figure 5.95 Liquid chromatograms of 10 μL samples of diluted incubation mixtures (A, blank incubation; B, assay incubation). After the 30 min incubation period, dilution with acetonitrile was such that the B*a*P concentration in the control flask (A) was 1 μ*M*; the same dilution was applied to the assay flask (B). The difference in the response gives the hydroxylation rate. (From Tulliez and Durand, 1981.)

Chrosorb RP 18) with a mobile phase of 10% water in acetonitrile. The column was eluted isocratically, and the B*a*P was determined with a fluorometer using a 366 nm excitation and monitoring emission at 385 nm.

The reaction mixture contained Tris-HCl buffer (pH 7.4), glucose-6-phosphate, NADP, glucose-6-phosphate dehydrogenase, and the microsomes. The mixture was preincubated, and the reaction was started by the addition of B*a*P dissolved in acetone. The mixture was incubated for 30 min at 37°C and terminated by the addition of cold acetone. Samples were diluted and injected for analysis. An example of the assay results is shown in Fig. 5.95.

The activity was from rat microsomes.

5.10.4 Hepatic Microsomal Testosterone Hydroxylase (van der Hoeven, 1984)

Testosterone has been used as a model substrate for different cytochrome P-450 monooxygenase activities. As a result of these oxidation reactions, multiple chemically related products are formed. In the assay developed for this activity, seven products are distinguished.

The separation of the substrate testosterone from the numerous reaction products was accomplished by reverse-phase hplc on a C-18 column preceded by a guard column. The compounds were separated by a combined isocratic and gradient elution procedure. The eluent was monitored at 240 nm. The separations obtained with these hplc conditions are shown in Fig. 5.96.

The reaction mixture contained 0.05 *M* HEPES-Na buffer (pH 7.4), $MgCl_2$, microsomes, and 1 m*M* testosterone. The reaction was initiated by the addition of the NADPH-generating solution, incubated for 10 min at 37°C, and terminated (after the addition of the internal standard corticosterone in methanol) with dichlormethane. The dichlormethane extract was dried and analyzed by hplc. A representative chromatogram of the metabolites of testosterone is shown in Fig. 5.97, and the rates of formation of the different metabolites are shown in Fig. 5.98.

The microsomal hydroxylase was prepared from male rats. Hepatic microsomes were isolated by treatment of a postmitochondrial fraction with polyethylene glycol 6000.

5.11 PTERIN METABOLISM

5.11.1 Folic Acid Cleaving Enzyme (DeWit et al., 1983)

This activity (FAS) cleaves folic acid into pterin-6-aldehyde and *p*-aminobenzoylglutamic acid, and the hplc assay developed for it involves the separation of the two.

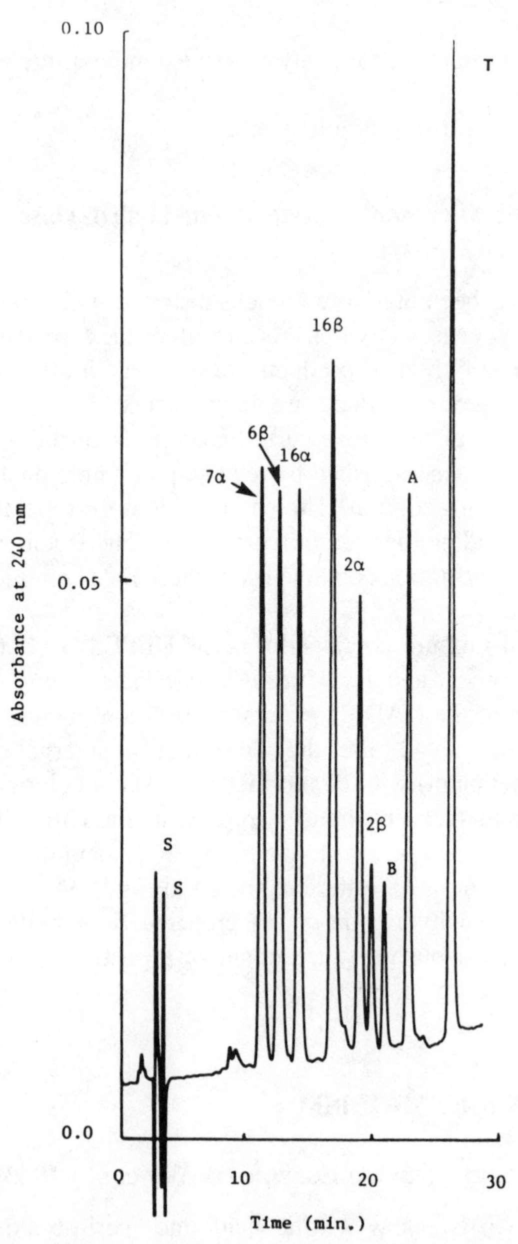

Figure 5.96 Separation of authentic standards of testosterone, its metabolites, and the internal standard corticosterone by hplc. S refers to solvent; 2α, 2β, 6β, 7α, 16α, and 16β are the hydroxylated metabolites of testosterone; A is androstenedione, B is corticosterone, and T is testosterone. (From van der Hoeven, 1984.)

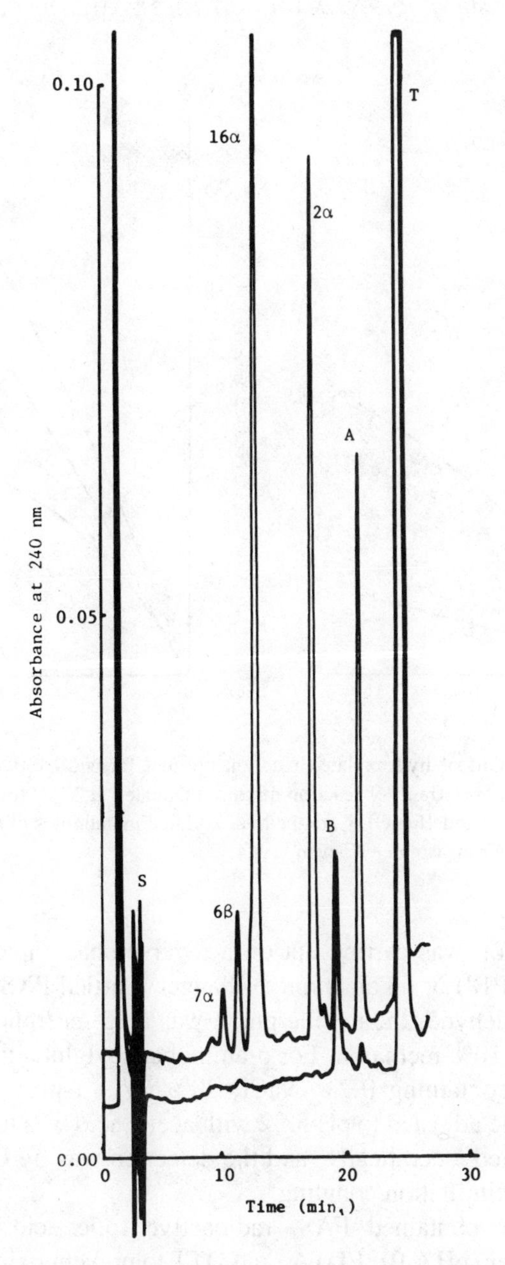

Figure 5.97 A typical chromatogram of the metabolites of testosterone formed by 1.0 mg of liver microsomal protein in 1.0 mL of reaction mixture incubated for 10 min at 37°C with (upper trace) and without (lower trace) a source of NADPH. (From van der Hoeven, 1984.)

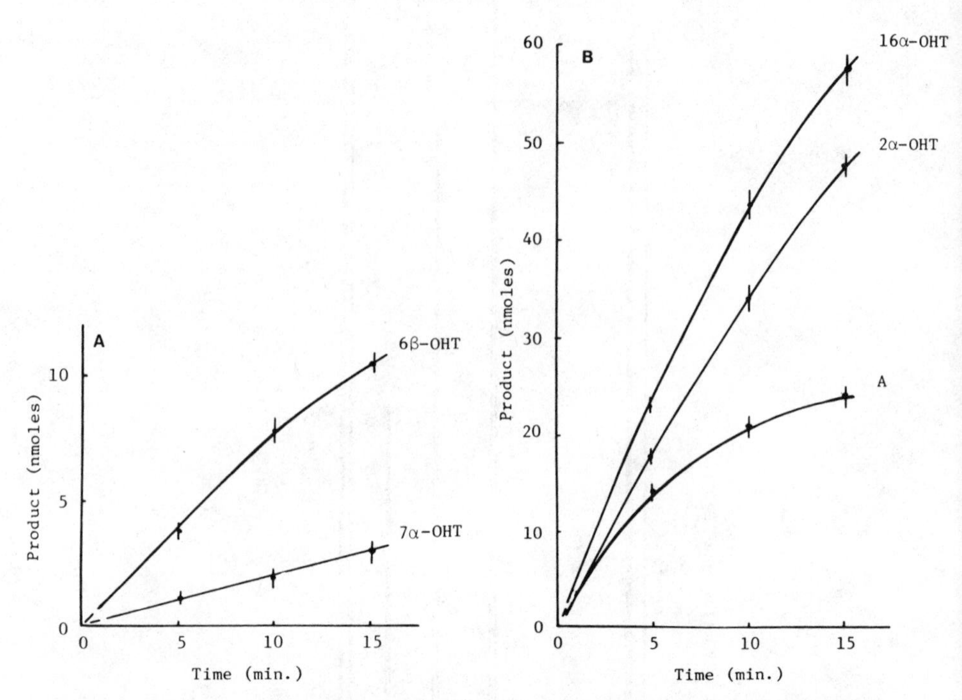

Figure 5.98 Amount of hydroxylated reaction product formed by 0.75 mg of hepatic microsomal protein in 1.0 mL of reaction mixture incubated at 37°C for different lengths of time. 2α, 6β, 7α, and 16α-OHT are the hydroxylated metabolites of testosterone. A is androstenedione. (From van der Hoeven, 1984.)

The separation was carried out on a reverse-phase hplc column (LiChrosorb 10 RP18) or on an anion exchanger (Partisil PXS 10/25 SAX). For pterin-6-aldehyde, the mobile phase was a 15 m*M* phosphate buffer (pH 6.0), with 10% methanol. For p-aminobenzoylglutamic acid a 0.1*M* NH_3 solution containing 0.2*M* NaCl, 20% (v/v) 1-propanol and 10% (v/v) acetonitrile adjusted to pH 5.32 with acetic acid was used. Both columns were eluted isocratically, and the detection was by UV at 254 nm and by liquid scintillation counting.

The mixture contained FAS, radioactive folic acid, a potassium phosphate buffer (pH 6.0), EDTA, and DTT to prevent oxidation of reaction products. The reaction was terminated by the addition of an ice-cold anion-exchange resin suspension in ammonium carbonate and ethanol.

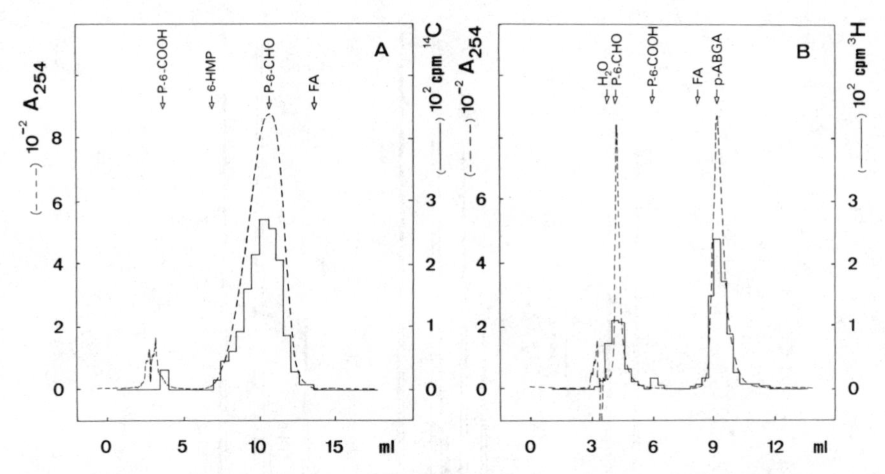

Figure 5.99 Identification of the products of folic acid C_9-N_{10} cleavage. (A) [2-^{14}C] folic acid was incubated with FAS and coinjected with pterin-6-aldehyde. Radioactivity was measured in 0.5 mL fractions. (B) [7,9,3′,5′-^{3}H]folic acid was incubated with FAS and coinjected with pterin-6-aldehyde and *p*-aminobenzoylglutamic acid. Radioactivity was determined in 0.5 mL fractions. Arrows indicate retention volumes of related compounds. Abbreviations: P-6-COOH, pterin-6-carboxylic acid; 6-HMP, 6-hydroxymethylpterin; P-6-CHO, pterin-6-aldehyde; FA, folic acid; p-ABGA, *p*-aminobenzoylglutamic acid. (From De Witt et al., 1983.)

Representative results are shown in Fig. 5.99 for the two chromatographic systems.

The enzyme activity was from the cellular slime mold *Dictyostelium minutum*.

5.11.2 Dihydrofolate Reductase (Reinhard et al., 1984)

Dihydrofolate reductase (EC 1.5.1.3) catalyzes the reduction of the 5,6 double bond in the H_2-folate to form H_4-folate. The activity also converts H_2-biopterin to H_4-biopterin, and the H_2-biopterin reductase may be the same enzyme. An hplc assay for the dihydrofolate reductase has been developed using H_2-biopterin as the substrate.

The separation of the substrate from the product was achieved by reverse-phase hplc using an ODS column. The mobile phase consisted of an

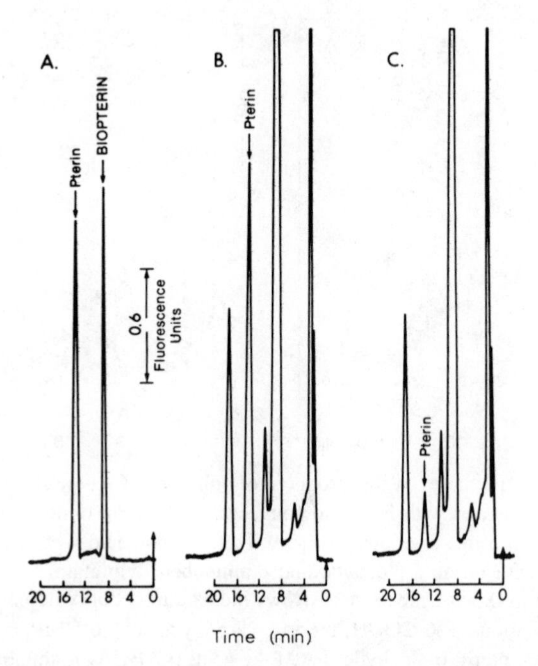

Figure 5.100 Chromatograms of enzyme activity. (A) Standard of 4 pmol biopterin and 10 pmol pterin; (B) brain extract incubated under assay conditions for 60 min; (C) boiled extract. The peak eluting at 18 min is imidazole. (From Reinhard et al., 1984.)

aqueous solution containing 0.5% acetonitrile and 0.1% tetrahydrofuran (v/v). The eluent was monitored fluorometrically with 350 and 450 nm excitation and emission wavelengths, respectively.

The assay mixture contained imidazole (pH 7.2), KCl, NADPH, glucose-6-phosphate, DTT, glucose-6-phosphate dehydrogenase, and H_2-biopterin. Samples were incubated for 60 min with tissue extract and terminated with 2 *N* TCA. After centrifugation (15,000 *g* for 5 min), the H_4-biopterin was oxidized to pterin with iodine, the reaction was terminated with ascorbic acid, and samples were injected onto the hplc column for analysis. An example of an assay is shown in Fig. 5.100, and a composite of the rate of product formation is shown in Fig. 5.101.

The activity was prepared from adult rat brain after homogenization, centrifugation, and desalting on G-25 Sephadex.

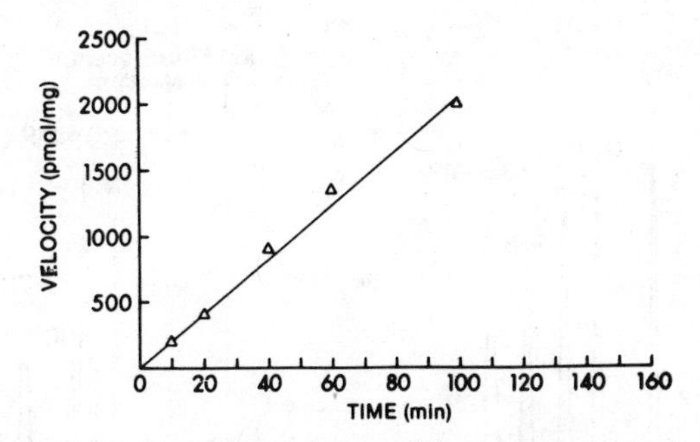

Figure 5.101 Effect of time on product formation. Desalted rat brain supernatants (510 μg of protein) were incubated for the times given on the ordinate. Boiled tissue blanks were subtracted. The concentration of H_2-biopterin was 250 μM. (From Reinhard et al., 1984.)

5.11.3 Guanosine Triphosphate Cyclohydrolase I (Blau and Niederwieser, 1983)

D-*Erythro*-7,8-dihydroneopterin triphosphate synthetase or GTP cyclohydrolase I (EC 3.5.4.16) catalyzes the formation of D-*erythro*-dihydroneopterin triphosphate (NH_2TP) from GTP. This activity is required for the synthesis of tetrahydrobiopterin. The hplc assay developed for this activity involves the direct measurement of neopterin phosphates after separation from GTP and its other hydrolytic products.

Separation was carried out by ion-paired reverse-phase hplc on a LiChrosorb RP-8 column with a mobile phase of isopropanol–triethylamine–85% phosphoric acid–water (3:10:3:984 v/v) at a final pH of 7.0. The column was eluted isocratically, and the nucleotides were detected at 254 and 287 nm and the pterins at 365 and 446 nm excitation and emission wavelengths, respectively. The separations obtained are shown in Fig. 5.102A and B.

The assay mixture contained GTP, EDTA, KCl, 10% glycerol, Tris-HCl (pH 7.8), and hydrolase. The reaction was incubated at 37°C for 90 min in the dark and was terminated by the addition of 1 *M* HCl containing 1% iodine and 2% KI. After 30 min, proteins were removed by centrifugation and excess iodine was removed with ascorbic acid. Samples were injected for analysis. Figure 5.103 shows the results of an assay, while Fig. 5.104 shows the activity in an homogenate as a function of incubation time.

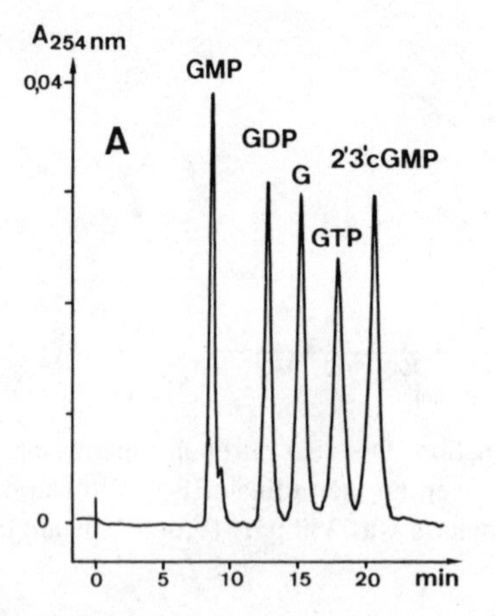

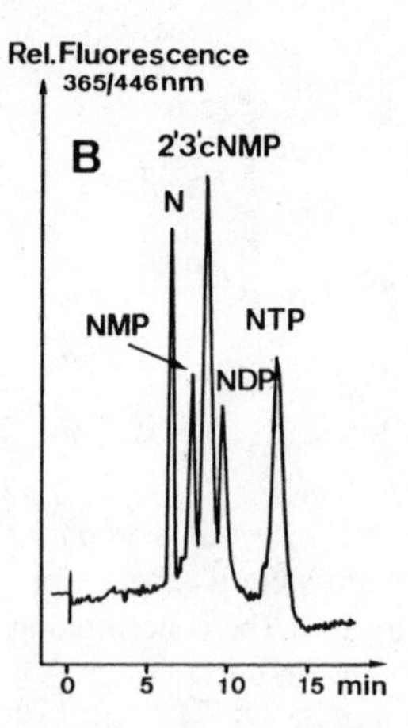

Figure 5.102 Hplc separation of (A) guanosine nucleotides and guanosine, injected amount 5–10 nmol each in 5 μL, and (B) neopterin phosphates. The mixture of neopterin phosphates injected was produced by partial hydrolysis of 16 pmol of NTP with alkaline phosphatase and addition of 2′,3′-cNMP. (From Blau and Niederwieser, 1983.)

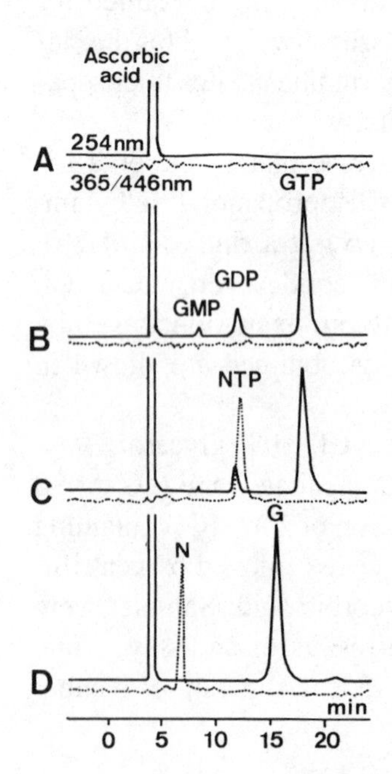

Figure 5.103 (A) Hplc of enzyme blank, mixture incubated without substrate; (B) substrate blank, mixture incubated without enzyme; (C) assay mixture; (D) assay mixture after treatment with alkaline phosphatase. Solid lines represent UV absorption at 254 nm. Dotted lines represent fluorescence at excitation = 365 nm, emission = 446 nm. (From Blau and Niederwieser, 1983.)

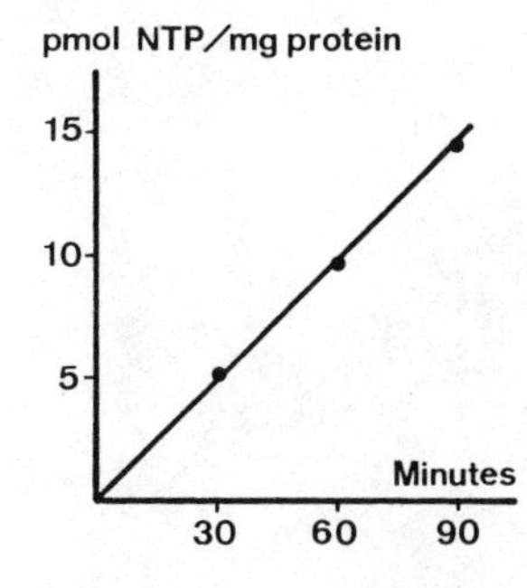

Figure 5.104 GTP cyclohydrolase I activity in rat liver homogenate as a function of incubation time. (From Blau and Niederwieser, 1983.)

The hydrolase activity was obtained from homogenized, rat liver, centrifuged, and applied to a Sephadex G-25 column. The eluate was used as the hydrolase.

5.12 SULFUR METABOLISM

5.12.1 Sulfotransferase (To and Wells, 1984)

Sulfotransferase catalyzes the transfer of sulfate from the donor molecule adenosine-3′-phosphate-5′-phosphosulfate (PAPS) to an acceptor, β-naphthol, to form the reaction product β-naphthol sulfate.

The assay used for this activity involves the separation of the β-naphthol from the product β-naphthol sulfate by reverse-phase hplc (C-18 column) with a mobile phase of 0.1 *M* acetic acid–acetonitrile (85:15 v/v). Absorbance was monitored at 235 nm.

The reaction mixture contained the substrate β-naphthol (0.125 *M*) in a phosphate buffer (pH 6.5), 5 m*M* 2-mercaptoethanol, and 0.2 m*M* PAPS in 5% acetone (v/v). The mixture was preincubated and the reaction started by the addition of a cytosolic fraction. The suspension was incubated for 10 min at 37°C, and the reaction was terminated by the addition of ice-cold methanol. After centrifugation to remove insoluble material, the supernatant solution was dried, redissolved in methanol, and analyzed by hplc.

Figure 5.105 shows the separation of the substrate from the product, and Fig. 5.106 shows the rate of formation of product during the incubation.

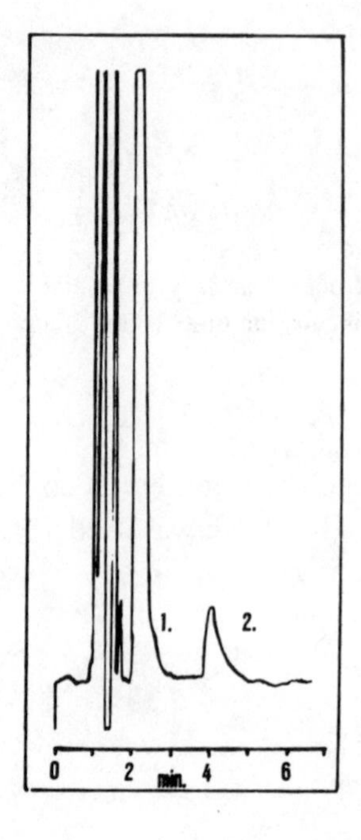

Figure 5.105 Chromatograms of the hplc separations of β-naphthol sulfate. Peaks: (1) acetaminophen, the internal standard; (2) β-naphthol sulfate. Solvent, 0.1 *M* acetic acid–acetonitrile (85:15, v/v); flow rate, 1.5 mL/min; wavelength, 235 nm. (From To and Wells, 1984.)

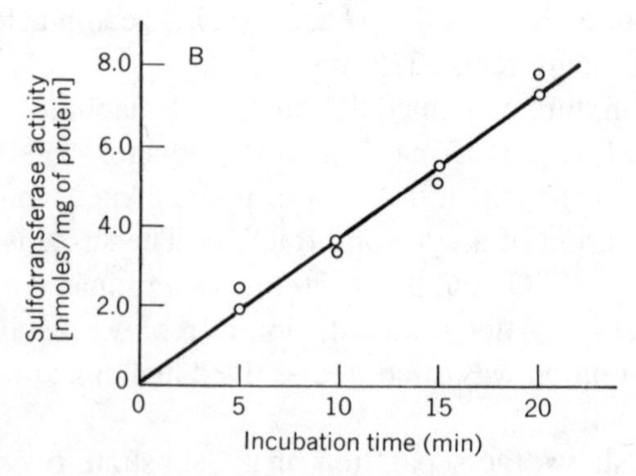

Figure 5.106 Effect of incubation time on enzyme activity. Sulfotransferase activity was measured by the amount of β-naphthol sulfate produced after varying incubation time periods. (From To and Wells, 1984.)

An hepatic cytosolic fraction obtained from mice was used as the source of activity.

5.12.2 Glutathione S-Transferase (Brown et al., 1982)

Glutathione S-transferase catalyzes the conjugation of products of cytochrome P450 metabolism to reduced glutathione (GSH). In the hplc method developed for this activity, styrene oxide is used as the substrate and the activity is measured by the formation of conjugates between the styrene oxide and the reduced glutathione conjugates.

The identification of the reaction conjugates was carried out by reverse-phase hplc on a C-18 (μBondapak) column eluted isocratically with a mobile phase of methanol–glacial acetic acid–water (20:1:79 v/v). The wavelength for detection was 254 nm.

The reaction mixture contained in 2 mL reduced glutathione (1 m*M*), sodium phosphate buffer (pH 7.8), and the transferase preparation. The reaction mix was preincubated at 37 °C, and the styrene oxide was added to start the reaction. Reactions were terminated by adding ice-cold ethyl acetate to extract the unreacted styrene oxide. The aqueous phase was dried and used for the hplc analysis.

Figure 5.107 shows a chromatogram of the reaction product. Two peaks were observed, and after analysis they were shown to be the conjugates expected.

The transferase was obtained from rat lung and liver cytosol fractions.

5.12.3 Adenosine 3′-Phosphate 5′-Sulfophosphate Sulfotransferase (Schwenn and Jender, 1980)

This sulfotransferase catalyzes the transfer of sulfate from adenosine 3′-phosphate 5′-sulfophosphate (PAPS) to an acceptor thiol.

All compounds, including substrates and both primary and secondary reaction products, were separated by ion-paired reverse-phase hplc (LiChrosorb RP-18) with a mobile phase of 9.4% 2-propanol containing tetrabutylammonium hydroxide at several pH values adjusted with phosphoric acid. The separations obtained are shown in Fig. 5.108. The column was eluted isocratically and monitored by UV at 254 nm and by liquid scintillation counting.

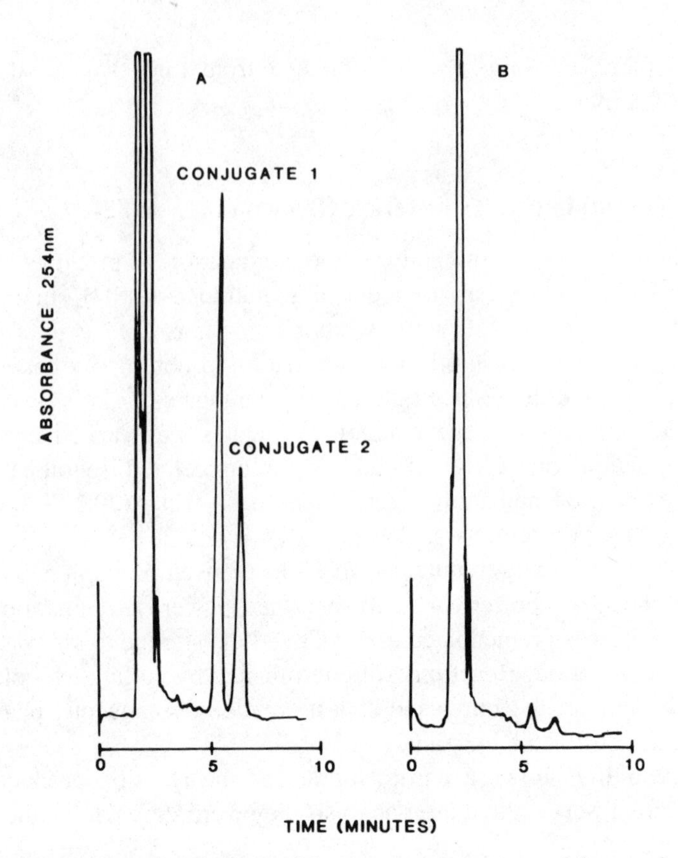

Figure 5.107 Hplc profile of 50 μL of the aqueous phase prepared from an incubation containing styrene oxide (1.0 m*M*), reduced glutathione (1.0 m*M*), and (A) 1.0 mg of native liver cytosol or (B) 1.0 mg of boiled liver cytosol. The UV absorbance was monitored at 254 nm. Detector sensitivities were 0.01 and 0.005 absorbance units full scale for A and B, respectively. (From Brown et al., 1982.)

The reaction mixture contained Tris-Cl (pH 8.0), $MgCl_2$, glutathione as the acceptor, [^{35}S]PAPS, and protein. Samples were withdrawn, and the reaction was terminated by forcing the sample through a microfilter at 6 bar under nitrogen. The filter retained the enzyme protein, and the nucleotides were recovered.

The activity was obtained from plants.

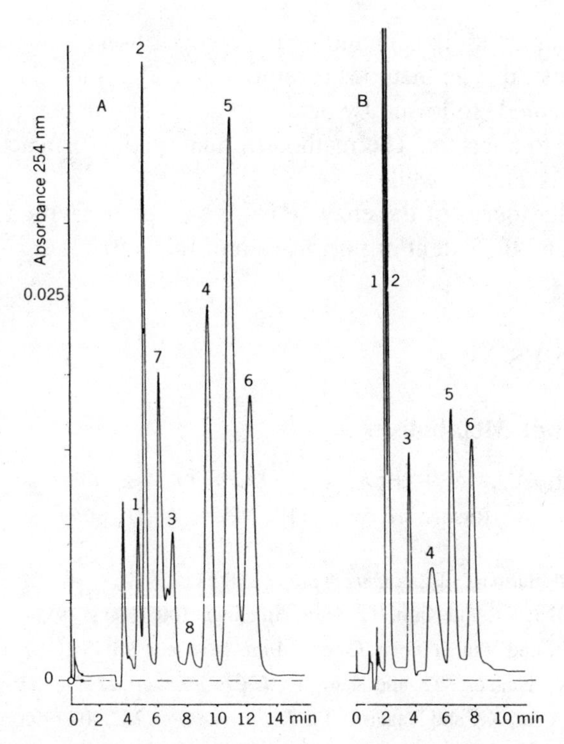

Figure 5.108 Separation of adenine nucleotides (mixture of authentic compounds) at pH 9.4 (A) and pH 8.0 (B). Column: Knauer RP-18/10, operated with a precolumn. Eluent: 9.4% 2-propanol, 3 m*M* TBAH, pH adjusted with 1 *N* phosphoric acid. Peaks: (1) 5′-AMP; (2) APS; (3) ADP; (4) ATP; (5) 3′5′-PAP; (6) PAPS; (7) NADP; (8) FAD. (From Schwenn and Jender, 1980.)

5.13 SUMMARY AND CONCLUSIONS

This chapter reviews hplc methods developed to assay enzyme activity. The enzymes have been grouped on the basis of their substrates, but in some cases an enzyme could have been put in several groups.

Each report has been reviewed according to a standardized format. The reaction is described first, and a general statement describes the basis for the assay. Next, the method of separation is described, including mention of the stationary phase, the composition of the mobile phase, and the method of elution of the column. The detection method is also described.

The reaction conditions including the composition of the reaction mixture are discussed. The methods for initiation and termination of the reaction are presented, followed by details concerning the preparation of the sample prior to analysis. The method of quantitation of reaction product is usually presented as well.

Finally, the source of the enzyme is given together with a synopsis of the method and the extent of purification of the activity.

REFERENCES

Catecholamine Metabolism

D'Erme, M., Rosei, M. A., Fiori, A., and Di Stazio, G., *Anal. Biochem.* **104:**59 (1980).

Feilchenfeld, N. B., Richter, H. W., and Waddell, W. H., *Anal. Biochem.* **122:**124 (1982).

Haavik, J., and Flatmark, T., *J. Chromatogr.* **198**:511 (1980).

Nagatsu, T., Oka, K., and Kato, T., *Anal. Biochem.* **100:**160 (1979).

Pennings, E. J., and Van Kempen, G. M., *Anal. Biochem.* **98:**452 (1979).

Rahman, M. K., Nagatsu, T., and Kato, T., *J. Chromatogr.* **221:**265 (1980).

Trocewicz, J., Oka, K., and Nagatsu, T., *J. Chromatogr.* **227:**407 (1982).

Proteinase

Advis, J. P., Krause, J. E., and McKelvy, J. F., *Anal. Biochem.* **125:**41 (1982).

Baranowski, R., Westenfelder, C., and Currie, B. L., *Anal. Biochem.* **121:**97 (1982).

Chen, C. S., Wu, S. H., and Wang, K. T., *J. Chromatogr.* **248**:451 (1982).

Gray, R. D., and Saneii, H. H., *Anal. Biochem.* **120:**339 (1982).

Horiuchi, M., Fujimura, K. I., Terashima, T., and Iso, T., *J. Chromatogr.* **233**:123 (1982).

Marceau, F., Drumheller, A., Gendreau, M., Lussier, A., and St. Pierre, S., *J. Chromatogr.* **266:**173 (1983).

Mousa, S., and Couri, D., *J. Chromatogr.* **267:**191 (1983).

Amino Acid Metabolism

Krstulovic, A. M., and Matzura, C., *J. Chromatogr.* **176:**217 (1979).

Martin, F., Suzuki, A., and Hirel, B., *Anal. Biochem.* **125:**24 (1982).

O'Donnell, J. J., Sandman, R. P., and Martin, S. R., *Anal. Biochem.* **90:**41 (1978).
Unnithan, S., Moraga, D. A., and Schuster, S. M. *Anal. Biochem.* **136:**195 (1984).

Polyamines

Haraguchi, R., Kai, M., Kohashi, K., and Ohkura, Y, *J. Chromatogr.* **202:**107 (1980).
Porta, R., Esposito, C., and Sellinger, O. Z., *J. Chromatogr.* **226:**208 (1981).

Heme Biosynthesis

Crowne, H., Lim, C. K., and Samson, D., *J. Chromatogr.* **223:**421 (1981).
Tikerpae, J., Samson, D., and Lim, C. K., *Clin. Chim. Acta* **113:**65 (1981).

Carbohydrate Metabolism

Fluharty, A. L., Glick, J. A., Samaan, G. F., and Kihara, H., *Anal. Biochem.* **121:**310 (1982).
Haegele, E. O., Schaich, E., Rauscher, E., Lehmann, P., and Grassl, M., *J. Chromatogr.* **223:**69 (1981).
Hymes, A. J. and Mullinax, F., *Anal. Biochem.* **139:**68 (1984).
Matsui, M., and Nagai, F., *Anal. Biochem.* **105:**141 (1980).
Naoi, M., and Yagi, K., *Anal. Biochem.* **116:**98 (1981).
Omichi, K.,and Ikenaka, T., *J. Chromatogr.* **230:**415 (1982).
Petrie, C. R., III, and Korytnyk, W., *Anal. Biochem.* **131:**153 (1983).
Sandman, R., *J. Chromatogr.* **272:**67 (1983).
Schreuder, H. A., and Welling, G. W., *J. Chromatogr.* **278:**275 (1983).
To, E. C. A., and Wells, P. G., *J Chromatogr* **301:**282 (1984).

Steroid Metabolism

Gallant, S., Bruckheimei, S. M., and Brownie, A. C., *Anal. Biochem.* **89:**196 (1978).
Suzuki, K., Kadowaki, A., and Tamaoki, B., *J. Endocrinol. Invest.* **4:**441 (1980).
Tanaka, Y., and Deluca, H. F. *Anal Biochem.* **110:**102 (1981).

Purine Metabolism

Ali, L. Z., and Sloan, D. L., *J. Biol. Chem.* **257:**1149 (1982).
Chism, G. W., Long, A. R., and Rolle, R., *J. Chromatogr.* **317:**263 (1984).

Danielson, N. D., and Huth, J. A., *J. Chromatogr.* **221:**39 (1980).

Dye, F., and Rossomando, E. F., *Biosci. Rep.* **2:**229 (1982).

Garrison, P. N., Roberson, G. M., Culver, C. A., and Barnes, L. D., *Biochemistry* **21:**6129 (1982).

Halfpenny, A. P., and Brown, P. R., *J. Chromatogr.* **199:**275 (1980).

Hanna, L., and Sloan, D. L., *Anal. Biochem.* **103:**230 (1980).

Hartwick, R., Jeffries, A., Krstulovic, A., and Brown, P. R., *J. Chromatogr. Sci.* **16:** 427 (1978).

Jahngen, E. G., and Rossomando, E. F., *Anal. Biochem.* **137:**493 (1984).

Pietta, P., Pace, M., and Menegus, F., *Anal. Biochem.* **131:**533 (1983).

Rossomando, E. F., Jahngen, J. H., and Eccleston, J. F., *Anal. Biochem.* **116:**80 (1981a).

Rossomando, E. F., Jahngen, J. H., and Eccleston, J. F., *Proc. Natl. Acad. Sci.* (*US*) **78:**2278 (1981b).

Rossomando, E. F., Cordis, G. A., and Markham, G. D., *Arch. Biochem. Biophys.* **220:**71 (1983).

Rossomando, E. F., and Jahngen, J. H., *J. Biol. Chem.* **258:**7653 (1983).

Sakai, T., Yanagihara, S., and Ushio, K., *J. Chromatogr.* **239:**717 (1982).

Tsukada, Y., Nagai, K., and Suda, H., *J. Neurochem.* **34:**1019 (1980).

Uberti, J., Lightbody, J. J., and Johnson, R. M., *Anal. Biochem.* **80:**1 (1977).

Oxygenations

Guenthner, T. M., Negishi, M., and Nebert, D. W., *Anal. Biochem.* **96:**201 (1979).

Richards, D. A., *J. Chromatogr.* **256:**71 (1983).

Tulliez, J. E., and Durand, E. F., *J. Chromatogr.* **219:**411 (1981).

van der Hoeven, T., *Anal. Biochem.* **138:**57 (1984).

Pterin Metabolism

Blau, N., and Niederwieser, A., *Anal. Biochem.* **128:**446 (1983).

De Wit, R. J. W., van der Velden, R. J., and Konijn, T. M., *J. Bact.* **154:**859 (1983).

Reinhard, J. F., Jr., Chao, J. Y., Smith, G. K., Duch, D. S., and Nichol, C. A., *Anal. Biochem.* **140:**548 (1984).

Sulfur Metabolism

Brown, D. L., Jr., Boda, W., Stone, M. P., and Buckpitt, A. R., *J. Chromatogr.* **231:** 265 (1982).

Schwenn, J. D., and Jender, H. G., *J. Chromatogr.* **193:**285 (1980).
To, E. C. A., and Wells, P. G., *J. Chromatogr.* **301:**282 (1984).

GENERAL REFERENCES

Catecholamine Metabolism

Allenmark, S., and Ali Qureshi, G., *J. Chromatogr.* **223:**188 (1981).

Boonyarat, D., Kojima, K., and Nagatsu, T., *J. Chromatogr.* **274:**331 (1983).

Fujita, K., Nagatsu, T., Maruta, K., Teradaira, R., Beppu, H., Tsuji, Y., and Kato, T., *Anal. Biochem.* **82:**130 (1977).

Koh, S., Arai, M., Kawai, S., and Okamato, J., *J. Chromatogr.* **226:**461 (1981).

Matsui, H., Kato, T., Yamamoto, C., Fujita, K., and Nagatsu, T., *J. Neurochem.* **37:** 289 (1981).

Nissinen, E., *Anal. Biochem.* **144:**247 (1985).

Nohta, H., Ohtsubo, K., Zaitsu, K., and Ohkura, Y., *J. Chromatogr.* **227:**415 (1982).

Zaitsu, K., Okada, Y., Nohta, H., Kohashi, K., and Ohkura, Y., *J. Chromatogr.* **211:** 129 (1981).

Amino Acid Metabolism

Blume, D., and Saunders, J. A., *Anal. Biochem.* **114:**97 (1981).

Hill, J. A., and Kitto, G. B., *J. Chromatogr.* **337:**397 (1985).

Imamura, I., Maeyama, K., Watanabe, T., and Wagde, H., *Anal. Biochem.* **139:**444 (1984).

Matsumoto, K., Imanari, T., and Amura, Z., *Chem. Pharm. Bull.* (*Tokyo*) **23:**1110 (1975).

Rahman, M. K., Nagatsu, T., and Kato, T., *J. Chromatogr.* **221:**265 (1980).

Trocewicz, J., Oka, K., and Nagatsu, T., *J. Chromatogr.* **227:**407 (1982).

Tsuruta, Y., Ishida, S., Kchashi, R., and Ohkura, Y., *Chem. Pharm. Bull.* (*Tokyo*) **29:**3398 (1981).

Polyamines

Chabannes, B. E., Bidard, J. N., Sarda, N. N., and Cronenberger, L. A., *J. Chromatogr.* **170:**430 (1979).

Pietta, P., Calatroni, A., and Colombo, R., *J. Chromatogr.* **243:**123 (1982).

Heme Biosynthesis

Cantoni, L., Ruggieri, R., DalFiume, D., and Rizzardini, M., *J. Chromatogr.* **229:**311 (1982).

Francis, J. E., and Smith, A. G., *Anal. Biochem.* **138:**404 (1984).

Carbohydrate Metabolism

Iwase, H., Morinaga, T., Li, Y. T., and Li, S. C., *Anal. Biochem.* **113:**93 (1981).

Kiuchi, K., Mutoh, T., and Naoi, M., *Anal. Biochem.* **140:**146 (1984).

Liu, Z., and Franklin, M. R., *Anal. Biochem.* **142:**340 (1984).

Stahl, R. L., Liebes, L. F., Farber, C. M., and Silber, R., *Anal. Biochem.* **131:**341 (1983).

Purine Metabolism

Fairbanks, L. D., Goday, A., Morris, G. S., Brolsma, M. F. J., Simmonds, H. A., and Gibson, T., *J. Chromatogr.* **276:**427 (1983).

Hartwick, R., and Brown, P. R., *J. Chromatogr.* **126:**679 (1976).

Krstulovic, A. M., Hartwick, R. A., and Brown, P. R., *J. Chromatogr.* **163:**19 (1979).

Ronca-Testoni, S., and Lucacchini, A., *Ital. J. Biochem.* **30:**190 (1981).

Rylance, H. J., Wallace, R. C., and Nuki, G., *Clin. Chim. Acta* **121:**159 (1982).

Vasquez, B., and Bieber, A. L., *Anal. Biochem.* **84:**504 (1978).

Oxygenations

Bacliani, M., Felici, M., Luna, M., and Artemi, F., *Anal. Biochem.* **133:**275 (1983).

Pietta, P., Calatroni, A., and Pace, M., *J. Chromatogr.* **241:**409 (1982).

Yamazoe, Y., Kamataki, T., and Kato, R., *Anal. Biochem.* **111:**126 (1981).

Chapter Six

Multienzyme Systems

Overview

This chapter describes the use of the hplc method to assay the activity of several enzymes simultaneously. The examples given include a case where several different enzymes can use the same substrate and form the same product, a case where a single enzyme can use different substrates to form different products, and a case where two different activities using the same substrate form different products. In another application the use of the hplc method to study metabolic pathways is described through a series of reconstitution studies, and finally an example of the application of the hplc method to the anabolism of adenosine is presented.

6.1 INTRODUCTION

Many investigators use enzymatic activities in a variety of ways. These include using them to characterize the stage of growth or development of a cell, as an indicator of the capacity to perform a specific physiological function, and as a measure of gene function, for example, when screening mutants for the absence of an activity. For purposes such as these, it is often necessary to determine the enzyme activity without purifying it first. Thus, assays for this purpose are often carried out in intact cells or in cell-free lysates. One problem with attempting to measure an activity under such conditions is that the assay methods that are usually used were designed to measure only a single component, either the substrate or the product, during the course of a reaction. Clearly, such assay methods make it difficult to quantitate the activity of one enzyme in the presence of others that can compete for the substrate or otherwise consume the product.

Since the hplc assay method can measure several related components simultaneously, it is possible to monitor the activity of a single enzyme under such conditions. In addition, because many of the enzymes that process related compounds represent a "pathway," the hplc method makes it possible to study the "flow" of a compound through such a multienzyme complex. The application of the hplc assay method to the study of metabolic pathways will be explored in this chapter, and experimental results of some studies will be presented.

The studies undertaken to date have utilized several approaches. In one, the hplc assay method has been applied to the question of assaying the activity of two enzymes that can use the same substrate but form different products. In another, the hplc method was applied to a related question where two different enzymes can use the same substrate and form the same product. And finally, it was applied to the case where the same enzyme can use either of two substrates to form two different products. In a separate application, the so-called reconstitution approach, a multienzyme pathway is reconstructed by the addition of enzymes one after the other to form a multienzyme complex and to thereby reconstitute a "naturally occurring" multienzyme complex. And finally, using this approach, an intact naturally occurring multienzyme system was studied using the hplc method.

6.2 ASSAY OF TWO ACTIVITIES FORMING DIFFERENT PRODUCTS FROM THE SAME SUBSTRATE (Pahuja et al., 1981)

The first example is that of the utilization of glutamic acid by two different activities: glutamine synthetase and glutamic acid decarboxylase.

Glutamine synthetase and glutamic acid decarboxylase catalyze the conversion of glutamic acid to glutamine and γ-aminobutyric acid

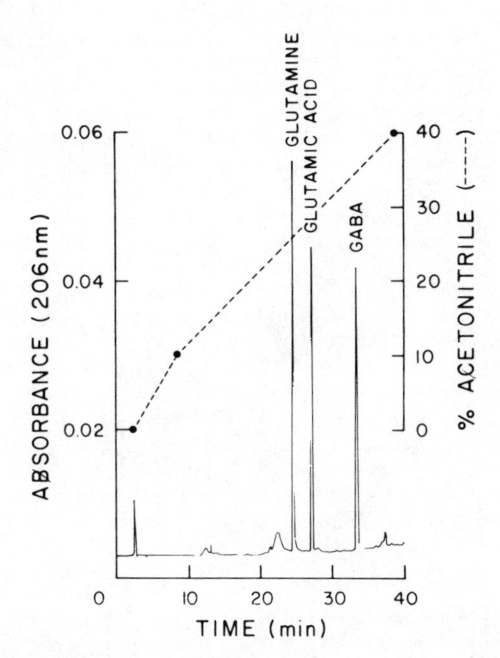

Figure 6.1 Chromatogram showing the separation of 0.5 μg of dansylated glutamine, glutamic acid, and GABA. The dashed line indicates the mobile phase gradient (% acetonitrile) used. (From Pahuja et al., 1981.)

(GABA), respectively. Both activities are involved in the metabolism of this amino acid, and it would be useful to be able to study both simultaneously.

An hplc method was developed to measure both activities in crude extracts.

The separation of glutamic acid and the two reaction products glutamine and GABA was accomplished by reverse-phase hplc (ODS-5 column) eluted with 100 m*M* potassium dihydrogenphosphate (pH 7.1) and a gradient of 0–40% acetonitrile. The compounds were derivatized with dansyl chloride, and the absorbance was monitored at 206 nm. Compounds were also detected by scintillation counting. Figure 6.1 shows the separation obtained using these conditions.

For the glutamine synthetase, the reaction mixture contained imidazole (pH 7.4), NH_4Cl, ATP, $MgCl_2$, and [^{3}H]glutamic acid. The reaction

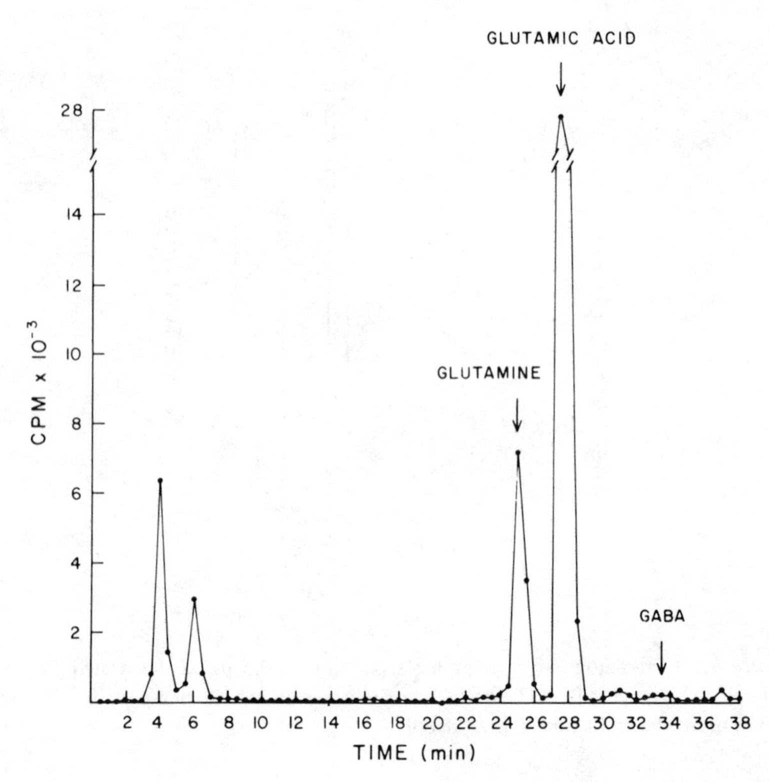

Figure 6.2 Chromatogram of the reaction mixture using 0.32 mg of crude brain extract. (From Pahuja et al., 1981.)

was started with the enzyme and terminated by cooling in an ice bath. After adding 1 *N* acetic acid, the insoluble material was removed by centrifugation and a sample removed for dansylation and analysis. The results of a synthetase assay are shown in Fig. 6.2. The formation of glutamine but no GABA is seen.

In contrast, when ATP is omitted and pyridoxal phosphate is added for the decarboxylase assay, the formation of GABA can be detected (Fig. 6.3).

The activities were prepared by homogenizing bovine brain, and after centrifugation at 1000 *g* to remove debris the supernatant solution was used directly for experiments. Bovine retinas were also prepared in the same manner.

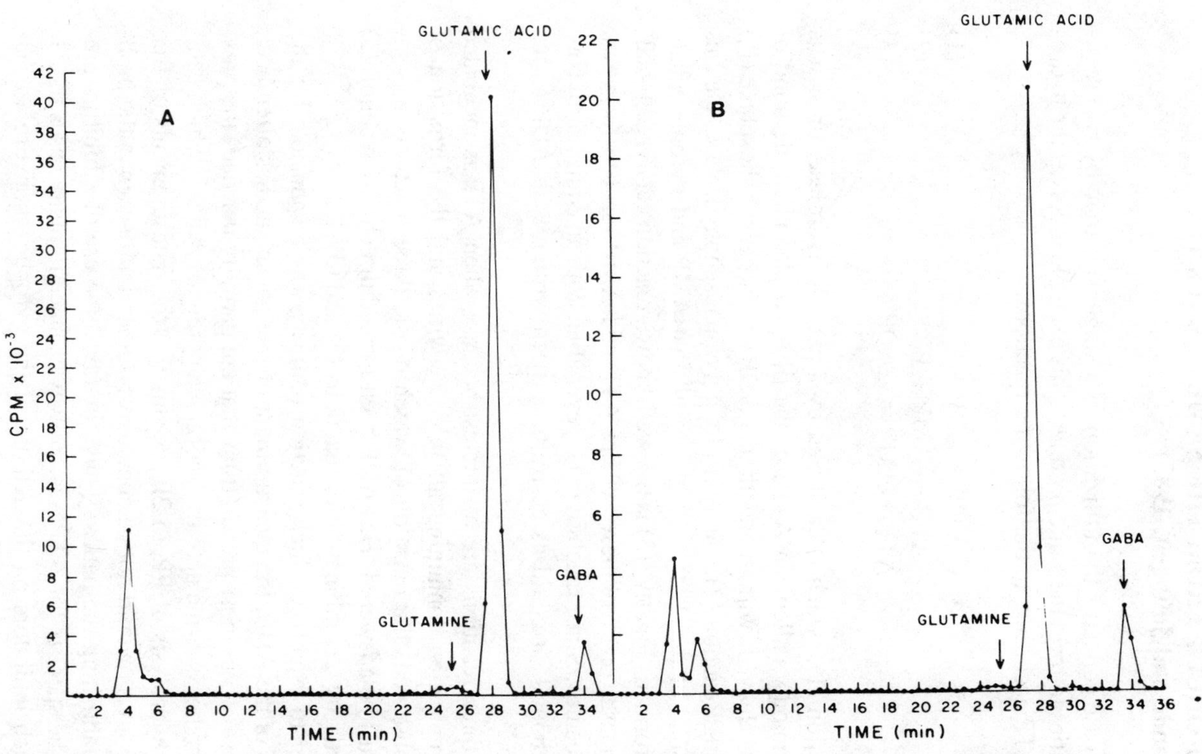

Figure 6.3 Chromatograms of the reaction mixture of (A) crude brain extract under GAD assay conditions; 0.17 mg of crude enzyme protein plus assay mixture was incubated for 1 h. (B) Crude retinal extract under GAD assay conditions; 0.30 mg of crude enzyme protein plus assay mixture was incubated for 2 h. (From Pahuja et al., 1981.)

6.3 ASSAY OF TWO ACTIVITIES FORMING THE SAME PRODUCT FROM THE SAME SUBSTRATE (Rossomando and Jahngen, 1983)

The next case concerns the utilization of a single substrate by two separate activities to form the same product. This example involves the hydrolysis of ATP according to the following reactions:

$$\text{ATP} \rightarrow \text{ADP} + \text{P}_i \quad (1)$$
$$\text{ATP} \rightarrow \text{AMP} + \text{PP}_i \quad (2)$$
$$\text{ATP} + \text{AMP} \leftrightharpoons 2\text{ADP} \quad (3)$$

In reaction (1), which is catalyzed by a phosphohydrolase, or as it is more commonly called, an ATPase, the β,γ-phosphoanhydride bond of the ATP is cleaved, with the result that ADP and inorganic phosphate (P_i) are formed. Reaction (2), catalyzed by a pyrophosphohydrolase, cleaves the α,β bond of the ATP, with the result that AMP and pyrophosphate (PP_i) are formed. Finally, (3) is the well-known reaction catalyzed by adenylate kinase, or myokinase, where ATP and AMP react to form two ADP molecules. Since reaction (3) is reversible, with an equilibrium constant of about 1, it could as easily have been written with ADP as the "substrate."

These three reactions are of interest because when ATP is incubated with a preparation containing all three activities and the formation of ADP is observed, the question of which of the three reactions was involved must be answered. Has ADP been formed directly by reaction (1) or indirectly by a combination of reactions (2) and (3)?

In order to establish which reaction pathway was responsible for the formation of the ADP, the measurement of free phosphate has often been used. However, in most preparations that are partially purified, the presence of another enzyme activity, inorganic pyrophosphatase, which catalyzes the hydrolysis of PP_i to $2P_i$, means that the measurements of free phosphate levels would not be conclusive. Other approaches to the problem of establishing the pathway have involved the use of inhibitors presumed to be specific for ATPases. However, since in many cases the effect of such inhibitors on the activity of the pyrophosphohydrolase has not been studied, some uncertainty creeps into the results of these studies as well. And finally, additional purification could always be carried out.

The hplc enzyme assay method provides an alternative procedure with which to approach this problem. In Chapter Five, hplc assays for these three activities were presented. Clearly, to be able to assay these activities by the hplc method, it is necessary to separate ATP, ADP, and AMP. As described in Chapter Five, this separation can be easily accomplished by ion-exchange hplc eluted isocratically with a mobile phase containing a phosphate buffer and sufficient concentration of salt to elute the ATP. Under these conditions, the order of elution of the compounds would be AMP first, ADP next, and ATP last.

To apply this method to the problem at hand, the AMP kinase activity should be measured first. Clearly, it would be advantageous to carry out this assay under conditions similar to those that would be present when only ATP was added to the complex. Thus, one might set up a reaction mixture with about 1 m*M* ATP and with AMP at nanomolar concentrations, that is, at a concentration that would be expected if the AMP had been derived from reaction (2). In addition, in order to follow its fate, in reaction (3) the AMP should be added to the incubation mixture in a radiolabeled form. Reaction (3) is started by the addition of the enzyme complex, and samples should be removed from the incubation mixture at suitable intervals, and injected onto the hplc column for analysis. After separation, the eluent should be monitored by both radiochemical and UV (254 nm) detectors.

If an AMP kinase is active, the following reactions would be expected to occur during the incubation:

$$\text{ATP} + {}^*\text{AMP} \rightarrow \text{ADP} + {}^*\text{ADP} \quad (4)$$
$$\text{ADP} + {}^*\text{ADP} \rightarrow \text{AMP} + {}^*\text{ATP} \quad (5)$$

where the asterisk indicates a radioactive compound. During the reaction, the products, radioactive *ATP and unlabeled AMP, would be formed. What about the ADP? Several routes are available for its formation, and one, of course, is reaction (4). However, two others can also occur. These are shown in reactions (6) and (7):

$$\text{ATP} + \text{AMP} \rightarrow 2\text{ADP} \quad (6)$$
$${}^*\text{ATP} + {}^*\text{AMP} \rightarrow 2{}^*\text{ADP} \quad (7)$$

Both reactions (6) and (7) would take place after reactions (4) and (5) had occurred and would utilize their products. For example, in reaction

(6), the unlabeled AMP formed in reaction (5) would react with unlabeled ATP that remained from the original substrate. As a result of reaction (6), two unlabeled ADP molecules would be formed, and therefore a small peak of unlabeled ADP would be expected to appear on the chromatogram.

In contrast, in reaction (7), the radioactive *ATP that was formed during reaction (5) would react with the radioactive *AMP that remained unreacted from reaction (4). However, detection of the reaction products of reaction (7) would be hindered by the excess unlabeled ATP still present in the reaction mixture, which would be expected to reduce the specific activity of any labeled *ATP formed during reaction (7), thereby making the formation of radioactive *ADP a low-probability event.

Finally, it should be noted that with a chromatographic system capable of separating ATP, ADP, and AMP and two monitors to detect both radiolabeled and unlabeled compounds, the presence of an ATPase would be readily detected, since such an activity would produce an excessive amount of unlabeled ADP with a corresponding reduction in the level of the unlabeled ATP. Of course, the availability of inhibitors of the myokinase such as P^1,P^5-diadenosine pentaphosphate makes it possible to test more directly the conclusion that the ADP formed is due to this activity, since as a result of its inhibition there would be an increase in the amount of AMP recovered and a proportional decline in the level of ATP.

Experiments have been carried out to demonstrate the method described above using a multienzyme complex. This complex was assayed first for the AMP kinase activity. A reaction mixture was prepared containing unlabeled ATP (1 m*M*) and radioactive [^{3}H]AMP only. The reaction was started by the addition of the complex, and samples were removed and analyzed by hplc. Chromatographic profiles, each representing the analysis of a sample removed from an incubation mixture at increasing times after the start of the incubation, are shown in Fig. 6.4. Both optical density and radioactivity were determined.

The profile of a sample taken from the incubation mixture early in the incubation shows two peaks, one of labeled *AMP with a retention time of 2 min and the other unlabeled ATP with a retention time of 9 min. No products have been formed. The next sample, obtained after a 10-min incubation, reveals the presence of small amounts of labeled *ADP, indicating that the AMP kinase is active [see reaction (3)], since an ATPase would, of course, produce unlabeled ADP [see reaction (1)]). Samples

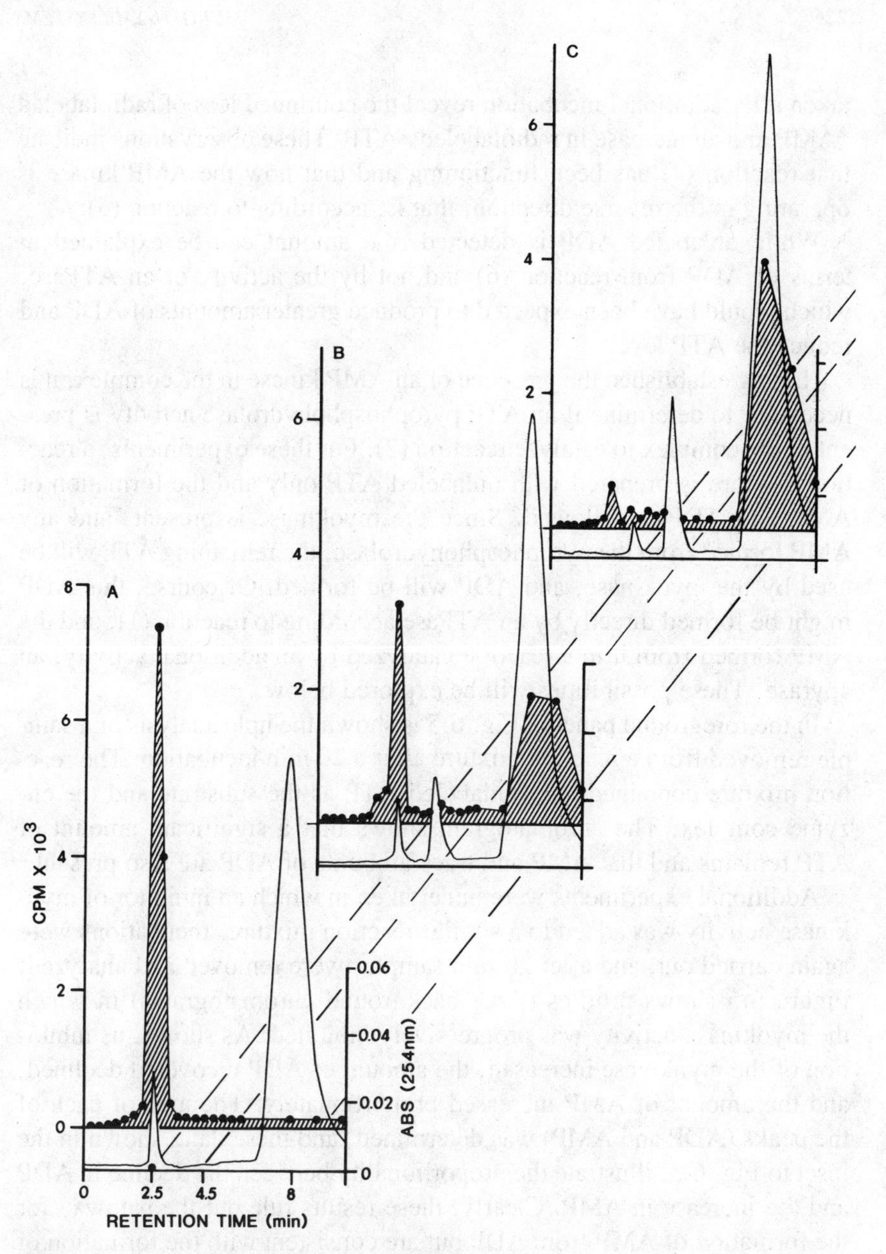

Figure 6.4 Adenylate kinase activity measured by the hplc assay method. The assay mixture contained 200 μ*M* ATP, 20 μ*M* [^{3}H]AMP (approximately 120,000 cpm), 50 m*M* Tris-HCl (pH 7.4), and 32 μg of protein of the enzyme from *D. discoideum*. Samples (20 μL) were injected, and the radioactivity was monitored with a Berthold LB 503 detector using PICO-Fluor 30 as the scintillant. Absorbance was measured at 254 nm.

taken after additional incubation reveal the continued loss of radiolabeled *AMP and an increase in radiolabeled *ATP. These observations indicate that reaction (4) has been functioning and that now the AMP kinase is operating in the reverse direction, that is, according to reaction (5).

While unlabeled ADP is detected, the amount can be explained in terms of ADP from reaction (6) and not by the activity of an ATPase, which would have been expected to produce greater amounts of ADP and reduce the ATP level.

Having established the presence of an AMP kinase in the complex, it is necessary to determine if an ATP pyrophosphohydrolase activity is present in the complex to catalyze reaction (2). For these experiments, a reaction mixture is prepared with unlabeled ATP only and the formation of AMP and ADP is followed. Since the myokinase is present, and any AMP formed from the pyrophosphohydrolase, the remaining ATP will be used by the myokinase, and ADP will be formed. Of course, this ADP might be formed directly by an ATPase according to reaction (1), and the AMP formed from it in a reaction catalyzed by an additional activity, an apyrase. These possibilities will be explored below.

In the foreground panel of Fig. 6.5 is shown the hplc analysis of a sample removed from a reaction mixture after a 20 min incubation. The reaction mixture contained only unlabeled ATP as the substrate and the enzyme complex. The chromatogram shows that a significant amount of ATP remains and that AMP and trace amounts of ADP are also present.

Additional experiments were undertaken in which an inhibitor of myokinase activity was added to a similar reaction mixture. Incubations were again carried out, and after 20 min samples were removed and analyzed. Figure 6.5 shows profiles (three background chromatograms) in which the myokinase activity was progressively inhibited. As shown, as inhibition of the myokinase increased, the amount of ADP recovered declined, and the amount of AMP increased proportionately. The area of each of the peaks (ADP and AMP) was determined, and these data, shown in the inset to Fig. 6.5, illustrate the proportionality between the decline in ADP and the increase in AMP. Clearly, these results rule out the pathway for the formation of AMP from ADP but are consistent with the formation of ADP as a result of the combined actions of reactions (2) and (3). In addition, these data suggest that an ATPase activity is present and would account for the formation of the unlabeled ADP observed in the original experiment.

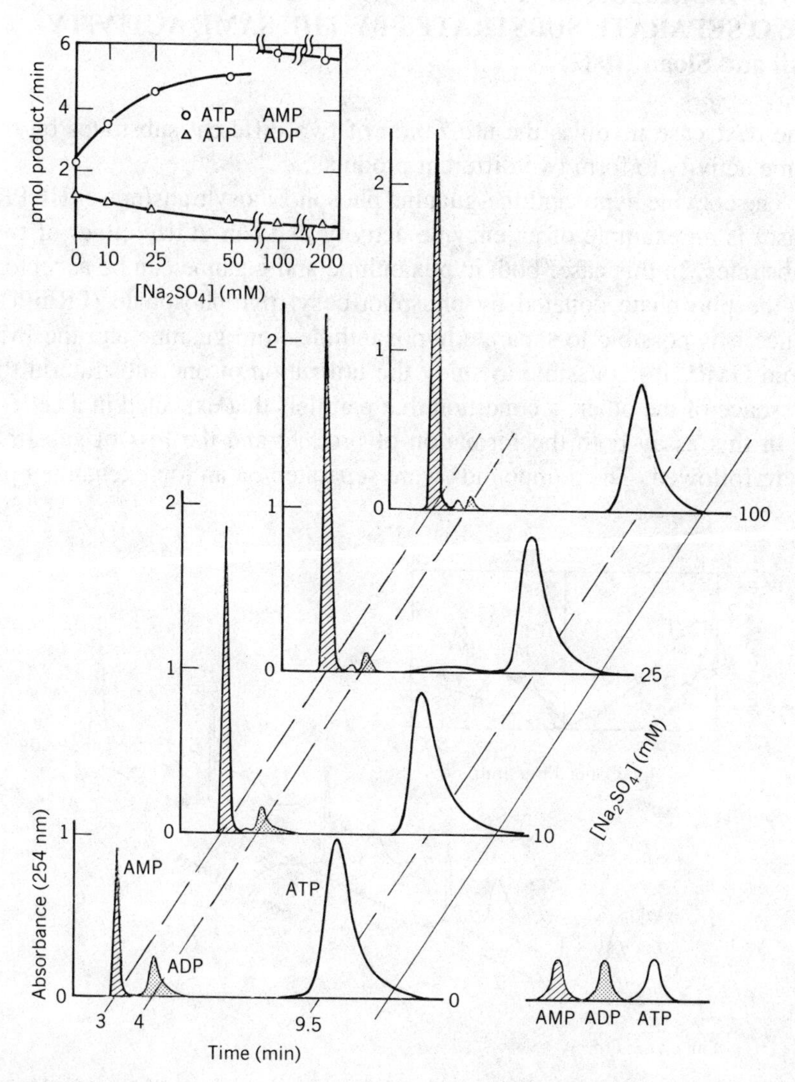

Figure 6.5 Effects of Na_2SO_4 on fate of ATP. A 30 μg sample of the extract was added to reaction mixtures containing, in a total volume of 200 μL, 100 μ*M* ATP, 0.4 m*M* Mn^{2+}, 50 m*M* Tris-HCl (pH 7.4), and 0–200 m*M* sodium sulfate. The reaction was incubated at 31 °C for 2 h and terminated by heating to 155 °C, and samples were analyzed by reverse-phase hplc. Chromatograms represent an analysis of each reaction mixture 2 h after the start of the reaction at four sulfate concentrations. Inset, initial rate of AMP and ADP formation from ATP. (From Rossomando and Jahngen, 1983.)

6.4 FORMATION OF TWO SEPARATE PRODUCTS FROM TWO SEPARATE SUBSTRATES BY THE SAME ACTIVITY (Ali and Sloan, 1982)

The next case involves the utilization of two different substrates by the same activity to form two different products.

The enzyme hypoxanthine-guanine phosphoribosyltransferase (HGPRTase) is an example of an enzyme activity that can utilize either of two substrates. In this case, both hypoxanthine and guanine can be acceptors of the phosphate donated by phosphoribosyl pyrophosphate (PRibPP). Since it is possible to separate hypoxanthine from guanine and the IMP from GMP, it is possible to study the utilization of one substrate in the presence of the other, a condition that parallels that expected in a cell.

In this assay both the formation of product and the loss of substrate were followed. The compounds were separated on an ion-exchange hplc

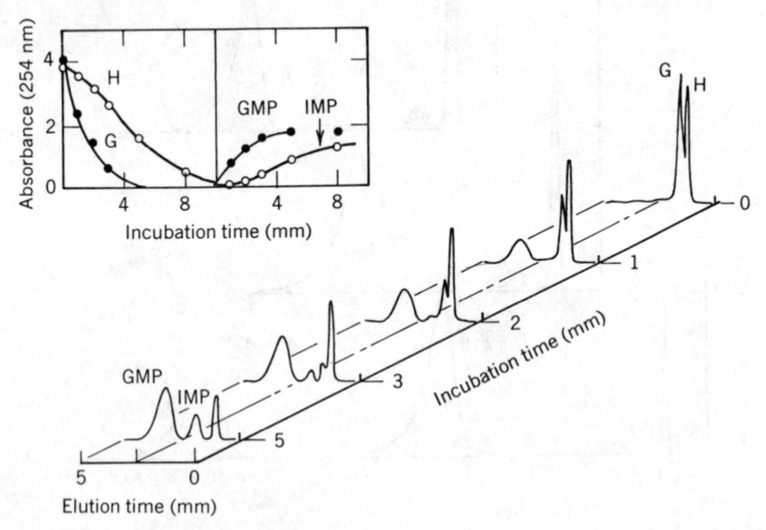

Figure 6.6 Hplc elution profiles of an incubation mixture made up of 2 nmol of hypoxanthine-guanine phosphoribosyltransferase, 50 μ*M* guanine (G), 50 μ*M* hypoxanthine (H), 100 μ*M* PRibPP, and 1 m*M* $MgCl_2$ in potassium phosphate (pH 7.4). At time intervals of 0–5 min, aliquots of the mixture were injected onto the hplc ion-exchange column and eluted. The inset illustrates the time-dependent utilization of H and G and formation of GMP and IMP as determined by the absorbance of each peak at 254 nm. (From Ali and Sloan, 1982.)

column equilibrated with 0.5 *M* phosphate buffer of pH 4.0. The eluent was monitored at 254 nm.

The reaction mixture contained 50 μ*M* guanine, 50 μ*M* hypoxanthine, 100 μ*M* PRibPP, and 1 m*M* $MgCl_2$ in potassium phosphate (pH 7.4). The reaction was initiated by the addition of the HGPRTase activity. At intervals the reaction was terminated by heating in a boiling water bath for 1 min. Denatured protein was removed by centrifugation, and the sample was purified further by filtration through a 0.45 μm type HA Millipore filter and injected for analysis.

The results of an experiment are shown in Fig. 6.6. The formation of GMP and IMP from guanine and hypoxanthine, respectively, can be followed. With this method it was possible to follow the initial rates of formation of IMP and GMP separately, the initial rates of both determined simultaneously, and the rate of PRibPP utilization with a fixed ratio of hypoxanthine and guanine.

The HGPRTase activity was purified from yeast.

6.5 ASSAY OF A MULTIENZYME COMPLEX BY THE RECONSTITUTION METHOD (Jahngen and Rossomando, 1984)

6.5.1 The Salvage Pathway: The Formation and Fate of IMP

"Salvage pathway" is a useful term to refer to that collection of biochemical reactions whose transformations result in the phosphorylation of purines. As a consequence of this phosphorylation, purines are not secreted by cells but, in fact, are returned to the cellular metabolic pool. One of these salvage enzymes is hypoxanthine-guanine phosphoribosyltransferase (HGPRTase), an activity that catalyzes the transfer of phosphoribose from phosphoribosylpyrosphosphate (PRibPP) to hypoxanthine, forming IMP and PP_i. Following its formation, IMP can undergo several other reactions as illustrated in Fig. 6.7. As a prelude to studies on the salvage of hypoxanthine by an intact multienzyme complex, a number of reconstitution studies were undertaken using purified enzymes to assess the suitability of the hplc enzyme assay system to carry out such studies on an intact multienzyme complex.

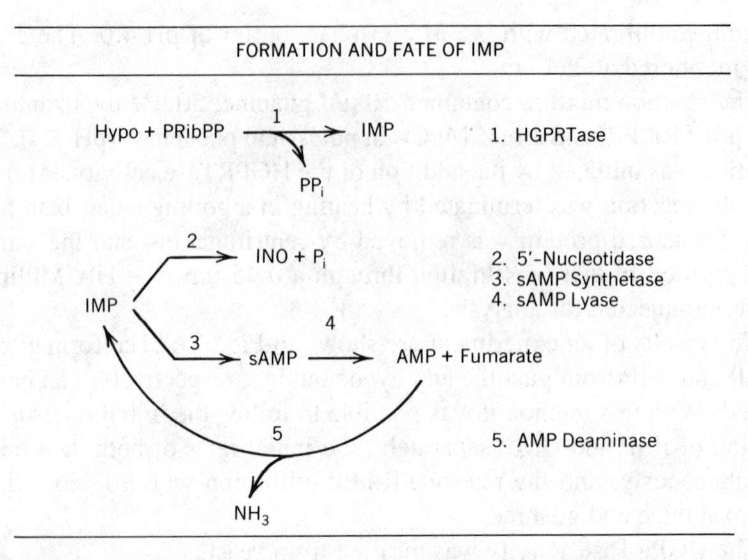

Figure 6.7 Schematic representation of the formation and fate of IMP. Formation of IMP is catalyzed by the enzyme hypoxanthine-guanine phosphoribosyl tranferase (1) from the substrate hypoxanthine (Hypo) and phosphoribosyl pyrophosphate (PRibPP). IMP is shown undergoing several reactions: the first (2) is catalyzed by 5′-nucleotidase to form inosine (INO) and orthophosphate (P_i); the other (3) is a two-step reaction catalyzed by sAMP synthetase to form adenylosuccinate (sAMP) and (4) by the enzyme sAMP lyase to convert sAMP to AMP and fumarate. Finally, (5) the deamination of AMP to IMP and NH_3 is catalyzed by AMP deaminase.

6.5.2 The Degradation of IMP to Inosine

The enzyme 5′-nucleotidase dephosphorylates IMP to inosine and P_i. Thus, as illustrated in Fig. 6.7, since this reaction represents a possible fate for the IMP formed by the transferase, reconstitution studies were undertaken with the nucleotidase. These studies were carried out using the hplc assay method developed for the HGPRTase activity as described in Chapter Five. A reaction mixture was prepared that contained hypoxanthine and PRibPP as substrates. The reaction was started by the addition of purified HGPRTase enzyme. Samples were removed and were analyzed by hplc. The chromatographic profiles obtained at zero, 10, 20, and 30 min after the start of the reaction, shown in Fig. 6.8, indicate the disappearance of hypoxanthine and the appearance of IMP, confirming the HGPRTase reaction.

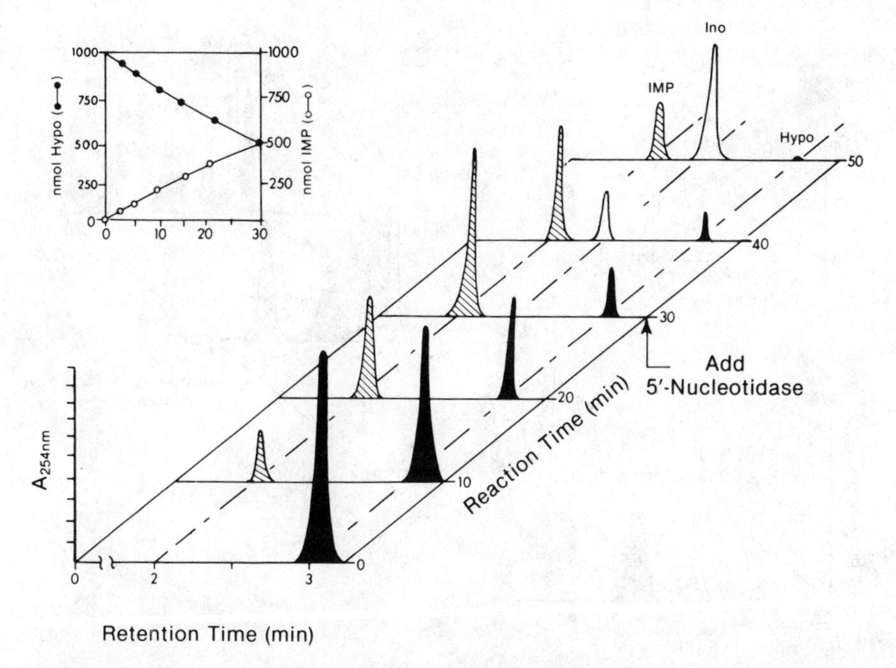

Figure 6.8 Hplc elution profiles of an incubation mixture to study hypoxanthine-guanine phosphoribosyltransferase activity. The reaction was initiated by the addition of the enzyme mixture, and aliquots were injected onto the hplc column at 10-min intervals. The solid peaks represent hypoxanthine, and the hatched peaks IMP. At 30 min, a 5′-nucleotidase activity was added to the reaction mixture. (From Jahngen and Rossomando, 1984.)

Immediately after removal of the 30 min sample, the 5′-nucleotidase enzyme was added to the incubation tube. The incubation was continued, and samples were again removed for analysis. The hplc profile in Fig. 6.8, obtained from a sample removed after an additional 10 min of incubation, shows an increase in inosine, the product of the 5′-nucleotidase activity, as well as a decline in IMP. Hypoxanthine continues to decline after the reconstitution of the system.

6.5.3 The Conversion of IMP to AMP

Dephosphorylation is not the only fate of IMP. An alternative fate is the formation of AMP by two reactions also illustrated in Fig. 6.7. The first reaction involves the condensation of IMP with aspartic acid to form the compound adenylosuccinate (sAMP). This reaction, catalyzed by the en-

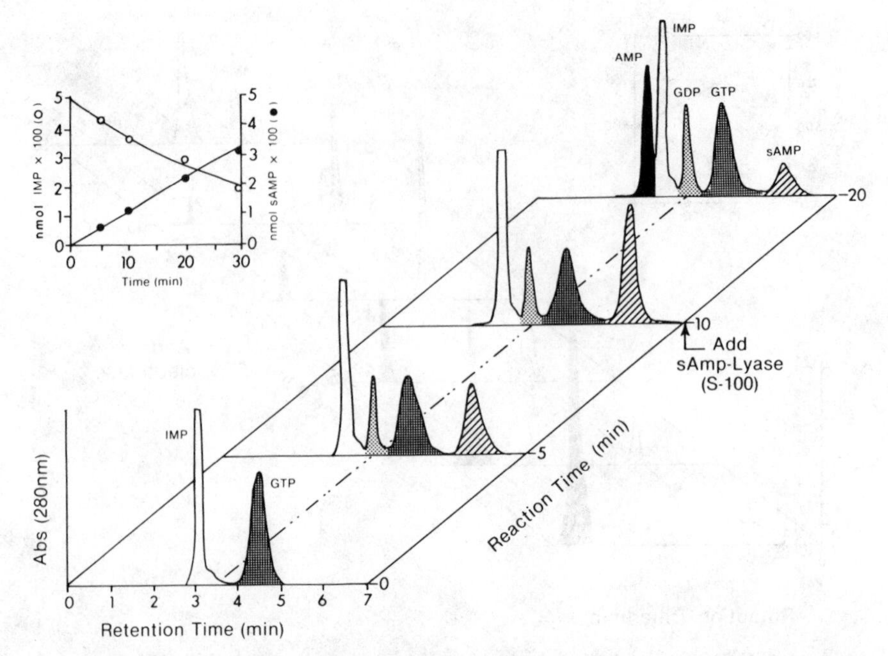

Figure 6.9 Hplc elution profiles of adenylosuccinate synthetase incubation mixtures. In this case the reaction was initiated by the addition of 1.25 μmol of aspartate (pH 7.4). At 5 min intervals, 20 μL samples were injected onto the hplc reverse-phase column and eluted. After 10 min of incubation (arrow), sAMP lyase was added to the incubation mixture along with 10 nmol of EDTA. The insert illustrates the time-dependent utilization of IMP and the formation of sAMP, as determined by integration of the respective peaks from the hplc chromatograms. (From Jahngen and Rossomando, 1984.)

zyme sAMP synthetase, is followed by another catalyzed by sAMP lyase, in which sAMP is cleaved to form AMP and fumaric acid. Since the hplc assay method can separate all these components (see Chapter Five), it would appear to be suitable for the study of such a multienzyme complex in which hypoxanthine could be salvaged to AMP. First, however, reconstitution studies were again carried out to test the capacity of the hplc assay method to follow just the processing of IMP to AMP.

A reaction mixture containing IMP, aspartic acid, and GTP was prepared, and the reaction was started by the addition of purified sAMP synthetase. As the incubation proceeded, samples were removed and analyzed by hplc. As shown in Fig. 6.9, the chromatograms showed the

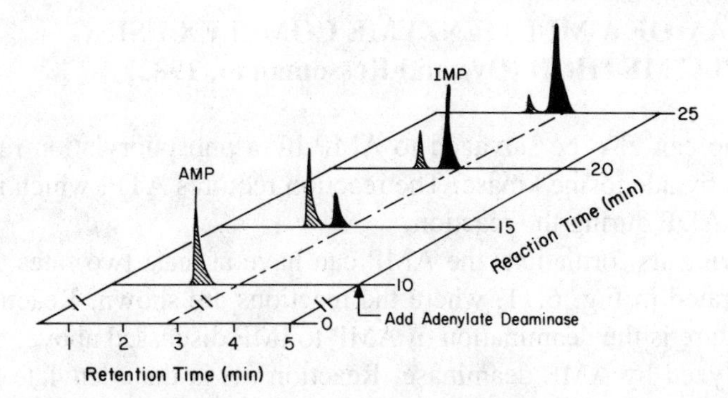

Figure 6.10 AMP was formed from adenosine and ATP in a reaction catalyzed by adenosine kinase. After a 10 min incubation, adenylate deaminase was added to the reaction mixture, and samples were taken and analyzed by hplc. Samples were analyzed every 5 min.

presence of the substrates and the formation of sAMP. After incubation for 20 min, the multienzyme system was reconstituted by the addition of a sample of sAMP lyase to the reaction. The reconstituted system was incubated for an additional 10 min, and a sample removed at that time was analyzed by hplc. The profile, shown in Fig. 6.9, illustrates a decline in the level of sAMP and the appearance of a new peak, AMP, confirming the successful reconstitution of this two-enzyme system.

6.5.4 The Return of AMP to IMP

While the formation of AMP from hypoxanthine is surely a good start, the salvage will be truly successful only if the AMP is converted to ATP. However, prior to continued phosphorylation, the AMP formed by the two-enzyme reaction sequence described above can undergo another fate—its deamination to form IMP and ammonia (see Fig. 6.7). Since the hplc method is able to separate AMP and IMP, reconstitution experiments were again undertaken to determine if the hplc could follow this reaction as well. A reaction mixture was prepared, AMP formed, and an AMP deaminase added to the reaction mixture. Samples were again removed and, as shown in Fig. 6.10, the addition of the AMP deaminase resulted in the conversion of AMP to IMP. Thus, this reaction sequence could also be followed.

6.6 ASSAY OF A MULTIENZYME COMPLEX USING THE HPLC METHOD (Dye and Rossomando, 1982)

Adenosine can also be salvaged to AMP by a phosphorylation reaction catalyzed by adenosine kinase. The reaction requires ATP, which is converted to ADP during the reaction.

Following its formation, the AMP can have at least two fates. These are illustrated in Fig. 6.11, where the reactions are shown. Reaction (2) of this figure is the deamination of AMP to IMP discussed above, a reaction catalyzed by AMP deaminase. Reaction (4) is the adenylate kinase reaction in which ATP is involved and two ADP molecules are formed. This reaction was also discussed above.

To follow the salvage of adenosine to AMP and the subsequent fate of AMP, an in vitro multienzyme complex was prepared from rat liver. In addition, in order to follow it directly, the adenosine was replaced by a fluorescent analog FoA (see Chapter Three for the structure of the analog). The reactions for FoA are shown in Fig. 6.12. For these experiments, a reaction mixture containing FoA and ATP was prepared. The complex was added, and samples were removed for hplc analysis. The

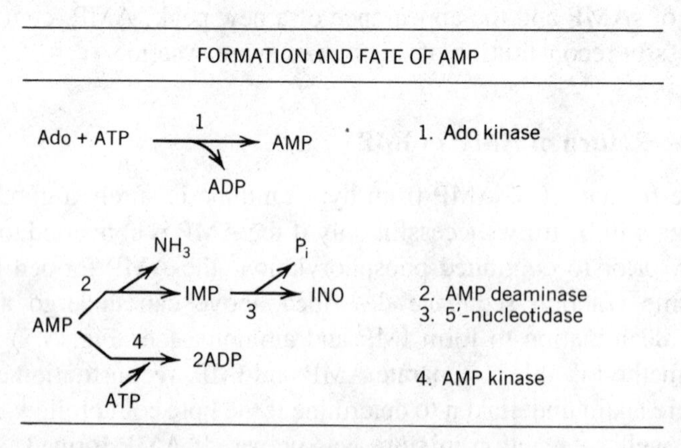

Figure 6.11 Representation of the formation and fate of AMP. AMP is shown formed by adenosine kinase (1) in a reaction which uses ATP as the phosphate donor and forms ADP as the second reaction product. AMP is shown being deaminated to IMP by the enzyme AMP deaminase (2) and converted to inosine (INO) by a 5′-nucleotidase activity (3). Also, AMP is shown being phosphorylated to ADP by the enzyme AMP kinase (4).

With Adenosine (Ado)		With Formycin A (FoA)
Ado + ATP → AMP + ADP	(1)	FoA + ATP → FoMP + ADP
AMP + ATP → 2ADP	(2)	FoMP + ATP → FoDP + ADP
2ADP → ATP + AMP	(3)	FoDP + ADP → FoTP + AMP

(a) FoA, FoMP, FoDP, FoTP, the formycin analogs of adenosine, AMP, ADP and ATP respectively.

(b) Reaction (1), (2) and (3) refer the reactions catalyzed by adenosine kinase, adenylate kinase in the forward direction and adenylate kinase in the reverse reaction.

(c) The reactions shown with adenosine (Ado) as these substrates are on the left and those with formycin A (FoA) on the right.

Figure 6.12 A comparison of the reaction products formed by AMP kinase with both AMP and its formycin analog formycin 5′-monophosphate as substrates.

chromatograms of samples removed at early times show the substrates FoA and reveal the formation of FoMP by the adenosine kinase (Fig. 6.13). Samples removed after additional incubation revealed the formation of FoDP and eventually FoTP in the reaction mixture. The results indicate that the hplc system was able to monitor the salvage of adenosine to AMP and its subsequent metabolism to ADP and ATP by the AMP kinase present in the multienzyme complex.

6.7 SUMMARY AND CONCLUSIONS

This chapter presents the use of the hplc method to assay multiple enzyme activities and individual enzymes with more than one function. A number of examples are given to illustrate that the method can be used to assay two activities using the same substrate to form different products, two activities using the same substrates to form the same product, and one activity using two substrates to form two products. Also, the hplc method can be used to study an in vitro multienzyme complex reconstructed by the addition of pure enzymes and finally to study a naturally occurring multienzyme complex.

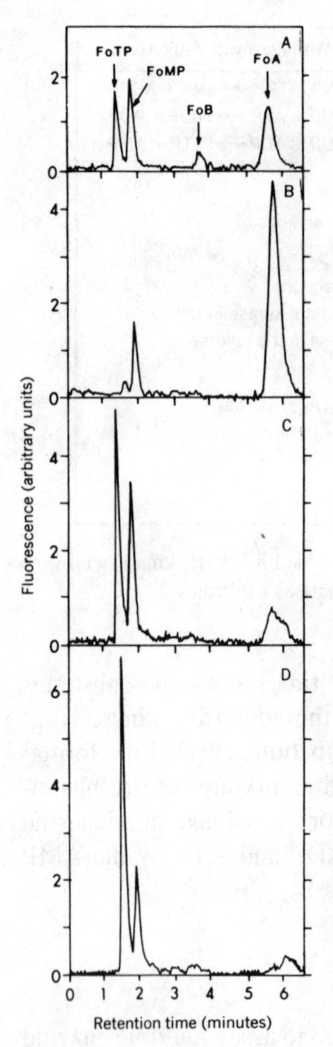

Figure 6.13 Hplc of FoA and its metabolites produced by activities present in an S-30 fraction prepared from rat liver. Operating conditions: Mobile phase, 0.1 *M* KH_2PO_4 adjusted to pH 5.5 with NaOH with 10% methanol; flow rate, 2 mL/min; room temperature; C-18 μBondapak column packing; fluorescence, excitation at 300 nm, emission about 320 nm. (A) Chromatogram obtained with standards as indicated at approximately 10 μg of each. Sensitivity of fluoresence detection, 0.5 μA units full scale. (B) Chromatogram obtained immediately after addition of S-30 to a reaction mixture containing FoA and ATP. (C, D) Chromatograms obtained after 8 and 16 min of incubation, respectively. FoB, formycin analog of inosine. (From Dye and Rossomando, 1982.) [Reprinted by permission from *Bioscience Reports,* **2**: 229–234 (1982).]

REFERENCES

Reconstitution Studies

Dye, F., and Rossomando, E. F., *Biosci. Rep.* **2:**229 (1982).

Jahngen, E. G., and Rossomando, E. F., *Anal. Biochem.* **137:**493 (1984).

Multiproduct Systems

Ali, L. Z., and Sloan, D. L., *J. Biol. Chem.* **257:**1149 (1982).
Pahuja, S. L., Albert, J., and Reid, T. W., *J. Chromatogr.* **225:**37 (1981).
Rossomando, E. F., and Jahngen, J. H., *J. Biol. Chem.* **258:**7653 (1983).

GENERAL REFERENCES

Multienzyme Systems

Friedrich, P., *Supramolecular Enzyme Organization*, Pergamon, Oxford, 1984.

Author Index

Subject Index